# The Institute of Refrigeration Proceedings 2022-2023

AA006177

Volume 119

2022-2023

ISBN: 978-1-7138-8038-7

**Printed from e-media with permission by:**

Curran Associates, Inc.
57 Morehouse Lane
Red Hook, NY 12571

**Some format issues inherent in the e-media version may also appear in this print version.**

Copyright© (2023) by The Institute of Refrigeration (IOR)
All rights reserved.

Printed with permission by Curran Associates, Inc. (2024)

For permission requests, please contact The Institute of Refrigeration (IOR)
at the address below.

The Institute of Refrigeration (IOR)
Kelvin House, 76 Mill Lane
Carshalton, Surrey SM5 2JR
Charitable Incorporated Organisation 1166869

Phone: +44 (0) 20 8647 7033

www.ior.org.uk

**Additional copies of this publication are available from:**

Curran Associates, Inc.
57 Morehouse Lane
Red Hook, NY 12571 USA
Phone: 845-758-0400
Fax: 845-758-2633
Email: curran@proceedings.com
Web: www.proceedings.com

IOR.org.uk

REFRIGERATION AIR CONDITIONING HEAT PUMPS

The knowledge hub for refrigeration, air conditioning and heat pumps

# The Institute of Refrigeration

# PROCEEDINGS 2022–23

# Volume 119

The Institute of Refrigeration (IOR) is an independent registered charity run for the public benefit. Its membership is over 2200 individuals whose work depends on refrigeration, air conditioning and heat pumps, providing technical information for designers, specifiers, installers, end users and service and maintenance personnel.

IOR publishes a range of Refrigerant Codes of Practice, Guidance Notes, Good Practice Guides and Safety Alerts. The ever growing library of up to date and reference publications contributes to making the IOR a central point of reference and resource for people for all matters relating to refrigeration, air conditioning and heat pump practice. IOR also offers events, webinars and e-learning. Visit www.ior.org.uk/membership for further information about our activities and how you can get involved.

# The Institute of Refrigeration

## PROCEEDINGS
## Volume 119
## 2022-23

---

## CONTENTS


Minutes of Meetings .................................................. 04

Annual Review 2021-22 .................................................. 06

*Celebrating 150 Years of Ammonia Compression*
Bernard Nagengast and Dr Andy Pearson FInstR .................................................. 14

*Presidential Address on the IOR's Part in the Net Zero Agenda for the Built Environment*
IOR President, Graeme Fox FInstR CEng .................................................. 34

*From Sustainable Cooling to Heat Recovery - Opportunities and Challenges for Waste Heat in Energy*
Prof Graeme Maidment PhD CEng FInstR .................................................. 48

*Is Money Best Spent on Efficient Design or Service in Refrigeration*
Robert Unsworth FInstR .................................................. 82

## Minutes of Meetings of Members 2022-2023

**3rd November 2022**

The IOR held a meeting to celebrate its formation. It was a joint event with the International Institute of Ammonia Refrigeration (IIAR) and was an online broadcast using Go to Webinar. 72 members and guests joined the live broadcast to hear Andy Pearson FInstR give his historical talk "150 years of Ammonia Compression - The Life and Times of David Boyle". The introduction was given by Eric Smith Vice President and Technical Director of the IIAR and the meeting was chaired by Dermot Cotter FInstR Chair of the IOR Papers and Publications Committee. The paper presentation was followed by a discussion that not only looked at the history of ammonia compression but also discussed the future of refrigerants. Questions and comments were sent in by Ian Fisher FInstR, David Blackhurst FInstR, Miriam Rodway MInstR, Robert Somerville and Loretta Britsch. The meeting closed at 5.11 pm.

**7th December 2022**

A meeting of 55 members and guests of the Institute of Refrigeration was held at Park Crescent Conference Centre in London to hear the Presidential Address given by the incoming President Graeme Fox. The paper was entitled "The IOR's Part in the Net Zero Agenda for the Built Environment". The meeting followed the IOR AGM, was broadcast using Go To Webinar and was chaired by Dermot Cotter FInstR, Chair of the Papers and Publications Committee. Questions were asked by Paul Singh FInstR, Steve Gill FInstR, Robert Lamb FInstR, Andrew Gigiel FInstR, Mike Creamer FInstR, Ian Fisher FInstR. Online questions were sent in by Neil Merrett, Andrew Fuller TMInstR and Ahmet Dermirtas and Martin Craxton. The meeting closed at 6.10 pm.

**2nd March 2023**

The March meeting took place as a virtual meeting and was broadcast using Go to Webinar. 97 members and guests join the live broadcast to hear the paper entitled "Opportunities and Challenges for Waste Heat in Energy Systems – From Sustainable Cooling to Heat Recovery" given by Graeme Maidment FInstR, winner of the J&E Hall Gold Medal. The meeting was chaired by Dermot Cotter FInstR, Chair of the IOR Papers and Publications Committee. The presentation was followed by a discussion. Questions and comments were submitted by Ian Bowman MInstR, Acar Cakici, Lisa-Jayne Cook FInstR, Martin Craxton, Declan Fitzmaurice, Nick Franzen MInstR, Tanja Groth, Alister Henderson, Moore Holmes, Doug Marriott FInstR, Will Overton MInstR, Phil Sheppard and John Thorne MInsR. The meeting concluded at 5.15 pm.

**13th April 2023**

The April presentation took place as a virtual meeting and was broadcast using Go to Webinar. 77 members and guests joined the webinar to hear the debate "Is Money Best Spent on Energy Efficient Design or Maintenance?" Robert Unsworth MInstR argued for the importance of design while Lisa Pogson represented the importance of maintenance. The meeting was chaired by Dermot Cotter FInstR Chair of the Papers and Publications Committee.

Questions and comments were submitted by John Bonner MInstR, Klas Berglöf MInstR, Valgard Bertschinger, Jose Augusto Castro Chagas, Graeme Fox FInstR, Jamie Furn, James Graham MInstR, Lawrence Leask MInstR, Doug Marriott FInstR, and Robert Somerville MInstR. The meeting closed at 5.10 pm.

## ANNUAL REVIEW 2021-22

### 1.1 Objectives and activities

The Trustees have reviewed these objectives and evaluated activities against them in line with the Charity Commission guidance on public benefit and confirm that the IOR continues to provide public benefit through:

- Activities to reduce the environmental impact of refrigeration, air conditioning and heat pump technologies by awareness raising how the sector can support end users to achieve net zero objectives and providing high quality technical events and publications to a wide audience primarily on line.
- Promoting training standards, education, skills and careers opportunities by engaging with students, colleges, schools and the Institute for Apprenticeships and Technical Education to revise apprenticeships and consider options for additional skills and training. We also promote diversity within the sector.
- Enhancing professionalism of individuals operating in the sector by offering ways of sharing information, encouraging Engineering Council professional registration, reviewing Apprenticeship Standards, recognising CPD (Continued Professional Development) of members, and promoting a Code of Professional Conduct as well as activities to promote equality and diversity.

The Trustees regularly review progress and emphasis of activities to achieve these aims at Board meetings. They confirmed that the following remain priorities for 2022:

1. Leadership in education – supporting apprenticeships
2. Promoting our industry to schools & young people
3. Setting the environmental agenda for net zero
4. Supporting members – students, technicians and young people grades

**The Impact and success** in progressing these strategic priorities is reviewed at quarterly Trustee meetings where progress reports are presented, and activity adjusted accordingly. In addition, Trustees monitor Engineering Council professional registration levels, international partnerships with Institutes sharing our objectives, opportunities to host international research conferences in the UK, local and subject based networking opportunities and providing public education and technical information on key issues such as employing apprentices, promoting the sector to schools, safety and environmental issues.

### 1.2 Signification activities and results

**Technical publications**

- A Guidance Note and Statement on Refrigerant selection.
- Technical papers on carbon dioxide hazard assessment, ammonia heat pumps for food and beverage industry, oil-injected screw compressors, history of hydrocarbon refrigeration.
- Service Engineer Guides published or updated on compressors, condensate pumps, evaporators, heat exchanger corrosion, brazing, recovery of refrigerant, alternative refrigerants, air conditioning technology.
- Conference Papers on Beyond Refrigeration Net Zero.
- Annual Proceedings Journal

**International profile**

- Participation in an ASHRAE international Associations Alliance meetings
- Continued collaboration with partners in Europe to promote REAL alternatives training and assessments.
- The IOR continued to support the UK membership of the International Institute of Refrigeration, an intergovernmental organisation
- An international conference on Sustainability and the Cold chain was in planning for April 2022.

**Networking and exchange of information**

- Free technical webinars were held on topics including working safely with flammable refrigerants, air conditioning efficiency, river source heat pumps,
- The Women in RACHP network continued to promote diversity with career development webinars and a practical training day held at Business Edge.
- SIRACH webinars on data centre waste heat recovery
- A short course on heat pump system efficiency was held on line
- Events were promoted from the IIR, Cold Chain Federation, Cryogenics Society, Institute of Refrigeration Ireland, ASHRAE, Refcold India, South West Society, Hampshire Society, and REAL Alternatives.
- Regular webinars introduced the benefits of membership, how to apply for Fellow and how to advance your membership grade. In addition networks on STEM promotion, Engineering Council registration and WinRACHP met regularly.
- Regular participation in ACRIB (Air Conditioning and Refrigeration Industry Board) meetings allowed IOR to contribute to the development of industry initiatives relating to education such as the ACRIB FGas Skillcard, to share technical information between industry associations and to monitor developments with legislation and policy in areas such as FGas Regulation, Building Safety Act and Building Regulations review.

**Education and Training**

- New technical guidance notes on Education were published on the topics of: Engineering Council registration, Using FantasticFridges website with schools, Employers guide to recruiting apprentices, skills for Design Engineers.
- Students on apprenticeship schemes were supported with free access to IOR publications and technical information on line.
- Winners of the IOR RAC Student of the Year competition were interviewed and articles published in an IOR page in the RAC Magazine.
- A network of STEM Ambassadors continued to meet to share ideas and resources
- A Steering Group of employers and trainers completed their review of the existing RACHP Engineering Technician Standard and Apprenticeship and this was submitted to Government.

**Encouraging and promoting innovation**

- Ted Perry Student Research Award was presented to Nausheen Basha of City, University of London for her work on screw compressor development.
- The J&E Hall International Gold Medal for practical contribution to the development of RACHP technology was not awarded during this period due to there being no Annual Dinner.
- The Beyond Refrigeration: Path to Net Zero initiative brought together expertise to support end users in their journey towards net zero and a group issued Policy Briefs on Heating and Cooling, and Education and Skills.

**The IOR benefits from significant contribution of volunteers, mainly members, through:**

- Membership of Committees and Working groups to edit publications, peer review technical papers and organise conferences or events
- Technical talks at webinars and other events
- Judging panels for awards and medals
- Reviewing and grading of membership applications

### 1.3 Achievements and performance

Significant activities generating income for the period included:

- Membership and Service Engineers Section fees and subscriptions
- Events
- Project grants

Success indicators were monitored and reported on at Trustee meetings included:

- Membership levels
- Attendance levels at webinars
- Numbers of publications produced and downloaded
- Website and social media reach
- Engineering Council registrations

The IOR is a professional affiliate of the Engineering Council, which allows its members to register with the Engineering Council through a licencing agreement with CIBSE. The Engineering Council provides valuable benchmarks against which the Trustees evaluate IOR activity and standards in areas such as Codes of Conduct, CPD, appeals procedures, diversity, corporate status, membership profile and marketing.

### 1.4 Fundraising

IOR does not engage in public fundraising or grant giving. It incurs expenditure related to membership marketing and manages awards for innovation, some of which are funded by donation by external organisations. IOR members (individuals) pay annual membership fees.

### 1.5 Note on impact of COVID-19

The IOR continued to make good use of online events and committee meetings during this period to ensure that services to members were not interrupted and to accommodate the ongoing risk of COVID-19 pandemic to staff, volunteers, members and service users. This resulted in an increase in people attending – particularly from overseas. It meant that costs were reduced (room hire, travel etc) and there was a corresponding reduction in income. Membership levels have remained steady during this period. All Staff continued to carry out their duties with hybrid working, some continuing remotely from home with all due security and GDPR processes in place, and most staff working in the office at least one day a week.

### 1.6 Future planned expenditure

Key activities being undertaken in the next period requiring investment of additional resources are:

- International conference on the Cold Chain planned for Spring 2022
- Maintenance of REAL Alternatives elearning outputs and expansion to more countries
- Supporting strategic priorities including Beyond Refrigeration Net Zero publicity and technical output, encouragement, and training of STEM ambassadors to give talks in schools.

## GOVERNANCE & MANAGEMENT

### 2.1 Legal Status

The IOR Constitution sets out the objects, powers, membership processes, decision making, AGM and voting arrangements, functions and duties of trustees including how trustees are appointed and inducted. The Rules set out the processes around grades of membership, membership fee collection and appointment of President, President Elect, Trustees, Hon Treasurer and Chief Executive. It also outlines the roles of Committees and Working Groups.

**Appointment of Trustees -** The Board of Trustees of the IOR includes six elected trustees, a President, President-Elect, Past President and Hon Treasurer. One-third of the elected Trustees retire from office each year by rotation. Vacancies are advertised by newsletter and on the website. Nominations must be supported by two members. If the number of nominations exceed the number of vacancies an election is held. New Trustees are given information about their legal responsibilities, terms of reference for Trustees, past meeting minutes, strategy documents and organisational, policy and financial background. They take part in an induction briefing with the President and Chief Executive. The President-Elect is voted in by the Membership usually one year before the current President is due to stand down. Presidents may serve a term of two-three years.

**Organisational structure -** The Board of Trustees appoints various Committees and Chair to carry out the work of the Institute as shown below. Each Committee includes a representative of the Trustees and is supported by a member of the permanent IOR staff. The IOR employs a permanent secretariat of five staff members led by a Chief Executive. Staff remuneration, including that of key staff, and the benefits packages for all staff, are reviewed annually by a Finance Committee comprised of Trustees in line with average cost of living increases and if appropriate exceptional staff performance.

**Decision making policies -** Trustees meet once a quarter for decision making purposes and may also make a resolution in writing or by email, by agreement of a simple majority. The Trustees have the power to delegate certain functions to a committee by determining the terms of reference. They can also authorise the formation of Branches. The Finance Committee is made up of a minimum of the Hon Treasurer, President and Immediate Past President and has authority for financial policies, monitoring of investment performance and resourcing including remuneration of staff based on merit and cost of living considerations. Policies are in place covering the reimbursement of Trustee expenses, Declaration of conflicts of interest, Diversity, Data Management, Risk management, Volunteer management, Code of Conduct and Discipline procedure and COVID working arrangements. Policies are reviewed on a regular rolling basis and updates made available to all Trustees.

**2.2 Trustees & Elections**

The members who served as trustees on 1st April 2020 and throughout the year to 31st March 2021 are shown below. The current President, Mike Creamer FInstR, is due to complete his term as President at the AGM in 2022.

Mike Creamer
President

Lisa-Jayne Cook
Elected Member

John Ellis
Elected Member

Ian Fisher
Elected Member

Graeme Fox
Elected Member

Kevin Glass
Past President

Juliet Loisell
Elected Member

Catarina Marques
Elected Member

Nick Rivers
Treasurer

John Skelton
Elected Member

At the AGM in November 2021 the term of office for the following elected members concluded and they stood down as elected members: Jacinta Caden and Graeme Fox. Graeme Fox was elected President-Elect and therefore remained on the Board. Two new members were elected: Ian Fisher and Catarina Marques.

**2.3 Committees**

- Annual Dinner – Juliet Loiselle (Chair), P Arrowsmith, R Hurley, Sam Buckell, Amy Gittoes.
- Education and Training – J Skelton (Chair), J Austin-Davies J Caden, S Buckell, R Mapperley, I Dickinson, D Russell.
- Finance – N Rivers (Chair), K Glass, M Creamer.
- International Refrigeration Committee – A Pearson (Chairman), M Camacho, J Evans, R Lawton, R Low, D Latta, J Miller, L McCusker, T Sekban and R Unsworth. The work of the Committee was supported by the following subscribing organisations: Cambridge Refrigeration Technology, GEA, Howden Compressors, Koura, J&E Hall International, Star Refrigeration, Spirax Sarco.
- Membership – D Wiszniewski (Chair), B Burton, A Thomas, S Kayes, A Riches, L-J Cook, C Marques, C Girdlestone, R Lamb.
- Papers and Publications – D Cotter (Chair), D Butler, G Hendra, K Glass, N Franzen, C Marques, C Vines, J Thompson, P Wilson, J Whitwell.
- RACHP Engineers' Section & ACHPI Management Panel – R Wright (Chair), I Fisher, D Garcia, P Stacey, J Broughton, G Adams, M Nankivell, G Wright, D Sowden, I Lemmon, M Carannante, B Parker, A Radford.
- Technical – D Paget (Chair), A Pearson, P Arrowsmith, S Benton, D Colbourne, J Evans, C Green, N Cook, N Atkins, N Roberts, N Rivers, P Cowan, P Holzer.
- Environment – G Maidment (Chair), B Churchyard, C Marques, C Green, LJ Cook, G Fox, T Mitchell, T Southall, R Lamb.
- Apprenticeship Steering Group – R Mapperley (Chair), J Austin-Davies, J Skelton, M Creamer and invited employers and trainers.
- Women in RACHP Network – K Perry (Chair), K Bullivant, J Caden, L-J Cook, C Fountain, J Loiselle, C Marques, A Prado, L Pogson, G Weston, S Gill.

### 2.4 Branch & Society Contacts

- East Anglian Refrigeration Society – C Sealy
- Northern Branch – M Morton
- IOR Scotland – D Watson
- London Refrigeration Society (affiliated) - B O'Reilly
- Hampshire Society (affiliated) – J Hobley
- Birmingham Society (affiliated) – P Singh
- SouthWest and South Wales Refrigeration Society (affiliated) - D Wiszniewski

### 2.5 External Representation & Co-operation

- Air Conditioning and Refrigeration Industry Board (IOR Directors who served during this period M Creamer, L-J Cook, J Caden)
- ASHRAE Associate Alliance (M Creamer, M Rodway)
- British Standards Institution and European and International Standards Review Committees (AB Pearson, D Colbourne, S Benton, D Paget, J Evans, M Rodway and others)
- International Institute of Refrigeration (UK delegate, J Evans)
- SIRACH (G Maidment, J Evans, L Waters)

### 2.6 Staff Members

- Miriam Rodway – Chief Executive
- Bawani Sarvendran – Accounts
- Clare Tooth – Membership
- Lisa Waters - Events Manager
- Symmone Cupidore – Publications

### 2.7 Professional advisors

- Auditor – Myrus Smith Chartered Accountants, Norman House, 8 Burnell Road, Sutton SM1 4BW
- Legal advisor – Third Sector Law, 28-30 High Street, Guildford GU1 3EL
- Bankers – CAF Bank, 25 Kings Hill Ave, Kings Hill, West Malling, Kent ME19 4JQ

## FINANCE & INVESTMENT POLICIES

The Trustees have prepared this report and results in the form of a Statement of Financial Activities in order to comply with the Statement of Recommended Practice for Charities.

The Trustees confirm that they have referred to the guidance contained in the Charity Commission's general guidance on public benefit when reviewing the Institute's aims and objectives and in planning future activities. The range of publications, events and projects provide benefit to the public by improving the effectiveness and efficiency of the refrigeration and air conditioning applications which underpin many aspects of public life. Specific activities and research such as supporting the implementation of new apprenticeships, raising awareness of careers and STEM Ambassadors' programmes, developing a network of trainers to improve safety and encourage the use of low GWP refrigerants, providing Guidance on the transition to Net Zero and encouraging innovation, safe practices and diversity. Activities and outputs are made widely available through a well visited website and promoted via trade press, newsletters, social media, and free webinars.

### 3.1 Financial performance

The Trustees budgeted cautiously this year due to ongoing disruption of the pandemic. A small deficit was forecast to allow for reinvestment of funds surplus to immediate requirements in strategic objectives to further the charity's benefit to the public such as education and technical services. The Annual Dinner that had been a fund raiser in previous years did not take place. Membership was forecast to remain static. The IOR maintains a diverse range of income sources including membership fees, events income, publication sales, and externally funded projects in order to mitigate risk of over-reliance on one income source.

The audited accounts for the year show an operating loss of £8,115 (2021 loss was £24,344) before investment revaluation. Investment values have continued to recover over the past two years and this year show an increase of £51,290 from the annual reappraisal of investment values. Therefore, there is a net increase of funds is £43,175 (compared to £68,458 for 2021).

The reconciliation of funds to be carried forward shows a total of £801,592 (2021 was £758,417) of which £442,982 are held in restricted funds that can only be used for specific purposes and £358,610 in the general fund whose use is unrestricted.

Business plans and financial forecast for the next five years aim to continue to invest reserves surplus to immediate requirements in order to provide for the short-term financial instability caused by the COVID-19 crisis and ensure longer term growth based on pursuing the strategic projects and services detailed in this report. Allocating resources to growing membership and retaining existing members through and leadership in education and technical areas remains a priority both to extend the Institute's range of benefits to the relevant community.

At the outset of the global pandemic in 2020 the IOR had in place a thorough and recently updated contingency plan and comprehensive risk assessment to accommodate the disruption and minimise the impact on its services. These policies and practices, updated, continued throughout the uncertainties of 2021. IT resources to allow staff to work remotely, paperless accounting, and investment in the latest professional level webinar facilities ensured committee meetings and events could continue uninterrupted. Whilst uncertainty remains the Trustees are assured that the IOR is well placed to respond flexibly to emerging market and business changes and to maintain a high level of services to members and the general public through web-based services. The Trustees have considered factors that could affect future financial performance which include reduced membership levels due to economic downturns, further drops in investment values, reduced opportunity to collaborate in funded projects and international physical events. The Trustees have in place cautious financial forecasts and have implemented plans to encourage targeted membership growth as well as continuing to monitor close investment portfolio performance.

### 3.2 Management of funds

Reserves Policy - The Trustees have established a reserves policy to minimise the risk to the Charity of unexpected downturns in income and allow for the development of new areas of activity to meet its long-term objectives. Trustees review this policy annually and consider that the appropriate level of reserves should be equivalent to six months operating expenditure, which is currently £164,363. It should be noted that this figure remains lower than pre-pandemic as fewer in person meetings and events were undertaken during the year and this situation is expected to change in the future with an Annual Dinner and Conference both planned for 2022-23 as in person events. The actual level of reserves calculated based on this year's accounts are £293,052 (unrestricted funds of £358,610 less tangible fixed assets of £65,558). It should be noted that approximately £ 52,125 of these reserves relate to cash balances held by IOR Branches and £2,922 are designated for future REAL Alternatives expenses. The current reserves held are therefore in excess of last years' six months expenditure however the Trustees consider that maintaining this level of reserves is prudent to manage future likely expenditure as operations return to normal levels post COVID-19 and to provide a mitigation for any further unforeseen circumstances during the current uncertain economic conditions.

The Trustees have judged to have sufficient funds to continue in operation for at least the next 12 months from the date of confirmation of the accounts, due to the high level of continued activity and engagement by members, the continued membership income levels, and availability of reserves upon which to draw should the need arise such as ongoing pandemic issues. The IOR budget forecasts are for a small deficit next year and to bring the operations in to balance in subsequent years.

Cash management - Cash surplus to immediate needs is invested prudently in line with recognised good practice for Charities, in a balanced mix of bank and building society deposit accounts CAF Bank (Charity Aid Foundation), Virgin Money Building Society, Cambridge and Counties Bank as well as in tax-free Charity investment bonds, which provide long term growth as well as substantial regular dividends. The IOR maintains a broad range of investments to minimise risk. Over the longer-term investment values have increased beyond that which would have been expected from simple bank-account based savings and these bonds continue to yield dividends. Trustees monitor the value of the investments through reports at each Board meeting.

### 3.3 Risk management

Trustees carry out a regular review of risks associated with the on-going management of the Charity and a risk register is held. Trustees are satisfied that processes and procedures are in place to mitigate the impact of such risks on the operations of the charity. The principal risks identified have been:

- Business disruption due to COVID-19 pandemic and economic downturn – leading to less events which in the past have provided substantial income, and a reduction in membership renewal. The impact of this situation is being monitored by Trustees with monthly updates. They are satisfied that IOR is well placed to ensure continued service for an indefinite period. This includes long term remote working for staff and organise of web-based meetings. An active programme of events is in place to maintain and increase membership going forward.
- Staffing – the Board review annually staff welfare, staffing requirements and opportunities to invest in IT to ensure adequate resources are available and that these are deployed efficiently with a business continuity plan to mitigate staffing related risks.
- Office working arrangements – efficient use is made of software and hardware to allow flexible working patterns, remote working, team working with volunteers in different locations, and integration of financial processes. An office COVID-19 risk assessment is continually updated to protect staff and visitors.
- Cybersecurity and data protection – IOR as a data owner keeps an up to date GDPR risk assessment, has processes in place to train staff, manage data sharing with processors, manage any potential breaches. It uses external consultants to ensure cybersecurity, updating of firewalls, prevention of security breaches etc.
- Membership profile – the need to attract more members to ensure a continued membership base in the long-term future is being addressed by improving relationships with training providers and providing networking and profile-raising opportunities for younger people. Activities are also targeting a broader range of refrigeration applications including air conditioning, heat pumps and heating to ensure a balanced representation of those working in the sector and contributors to technical output.
- Over-reliance on a small number of volunteers and balance of member input – the increasing technical and educational output needs an increased volunteer input with a more diverse range of experience. An annual recruitment drive to attract new member volunteers, specifying air conditioning or training experience, has been implemented. The IOR equality and diversity policy encourages equal opportunities and a balance of representation.
- Unpredictability of investment income and market revaluation of investments – remains a risk which the IOR addresses by ensuring a varied portfolio of charity investments, monitoring throughout the year their growth and returns and having a split of surplus funds between investments and bank/building society deposit accounts.

### 3.4 Designated and Restricted funds

Expenditure is made directly for the purposes for which these funds were established as shown in the notes to the accounts. Planned expenditure for the next period is as follows:

- Presidents' fund will continue to support from its income, education initiatives as considered most effective by the President and Board
- Ted Perry award will provide prizes for student researchers and to all student or apprentice winners identified by industry competitions
- Mike Tinsley fund to support a disabled engineer continues to be promoted widely
- The Funds representing EU grants for the CryoHub and REAL Alternatives 4 LIFE project have now been wound down.
- A Designated Fund for REAL Alternatives is holding funds raised by the project to cover costs of maintaining and developing future language versions of the elearning.

The financial statements that follow show the current values of the restricted, unrestricted and cash funds of the Institute.

## STATEMENT OF TRUSTEES' RESPONSIBILITIES

The trustees are responsible for preparing this Annual Report and the financial statements in accordance with applicable law and United Kingdom Accounting Standards (United Kingdom Generally Accepted Accounting Practice).

The law applicable to charities in England & Wales requires the trustees to prepare financial statements for each financial year which give a true and fair view of the state of affairs of the charity and of the incoming resources and application of resources of the charity for that period. In preparing these financial statements, the trustees are required to:

- select suitable accounting policies and then apply them consistently;
- observe the methods and principles in the Charities SORP 2019 (FRS 102);
- make judgements and estimates that are reasonable and prudent;
- state whether applicable accounting standards have been followed, subject to any material departures disclosed and explained in the financial statements;
- prepare the financial statements on the going concern basis unless it is inappropriate to presume that the charity will continue in operation.

The trustees are responsible for keeping proper accounting records that disclose with reasonable accuracy at any time the financial position of the charity and enable them to ensure that the financial statements comply with the Charities Act 2011, the Charities (Accounts and Reports) Regulations 2008 and the provisions of the trust deed. They are also responsible for safeguarding the assets of the charity and hence for taking reasonable steps for the prevention and detection of fraud and other irregularities.

Mike Creamer FInstR

President

Date 8th September 2022

IOR.org.uk

REFRIGERATION AIR CONDITIONING HEAT PUMPS

The knowledge hub for refrigeration, air conditioning and heat pumps

# Celebrating 150 years of ammonia compression

Bernard Nagengast and Dr Andy Pearson FInstR

Copyright © 2022
Institute of Refrigeration

## Why you should attend

1. Find out about the history of the early development of ammonia refrigeration.
2. Understand the contribution of some early pioneers of ammonia refrigeration during the 1800s–including David Boyle.
3. Celebrate over 123 years of the Institute of Refrigeration and 150th anniversary of the world's first ammonia refrigeration system using vapour compression.

## Introduction

Mechanical refrigeration using a vapour compression cycle was first demonstrated in London in 1834 and was very slow to develop into a viable commercial proposition but by the mid 1870s it started to take off. Over that initial period ammonia was suggested as a possible refrigerant several times but was not successfully demonstrated until the 1870s. It is reasonable to suggest that the successful adaptation of vapour compressors to the use of ammonia as the working fluid is linked to the commercial success of vapour compression as the preferred method of achieving mechanical refrigeration. There are several versions of the way in which ammonia was brought into the family of refrigerant working fluids. This paper aims to dig into the stories behind the superficial understanding of this development and give credit to some pioneers who may be less familiar to the modern audience.

Refrigerant expert Jim Calm has noted (Calm, 2012) that the history of refrigeration can be divided into four epochs and he characterises the first, from 1830 to 1930 as "whatever worked", meaning that the working fluids used in these systems were not developed for that purpose but were already available and could be readily applied to the refrigeration cycle. This rather underplays the significant technical developments that were required to make refrigeration systems safe, efficient and reliable. It might be more honest to describe most of the early developments as "whatever didn't work". Often progress was delayed because the mechanical difficulties of adapting existing available equipment to work in a refrigeration cycle. This appears to have been the case with ammonia, which was identified by William Cullen as the most favourable fluid for this purpose as early as 1755 and was listed as a possible refrigerant by many people. In his 400-page treatise *"Ammoniaque dans L'Industrie"*, Charles Tellier wrote *"Les différentes applications industrielles de l'ammoniaque, dont j'ai signalé la possibilité, ont assez vivement excité l'attention. De divers côtes, des demandes de renseignements m'ont été adressées. C'est pour répondre à la plupart d'entre elles, satisfaire l'empressement qui a accueille mes vues, que je publie cet ouvrage."* [The different industrial applications of ammonia, the possibility of which I have pointed out, have excited much attention. Enquiries have been made to me from various quarters. It is to answer most of them, to satisfy the eagerness which welcomed my views, that I publish this work.] (Tellier, 1867)

At the conclusion of the book he wrote *"Ai-je énoncé toutes les applications qu'il est possible de faire de ce corps? Non, assurément! Quelles qui soient les recherches tentées en vue d'applications nouvelles, j'applaudirai à leur succès, comme aussi toutes mes sympathies seront pour les travaux, qui, en vulgarisant ce corps à tant d'égards intéressant, favoriseront son entrée dans la pratique industrielle."* [Have I stated all the applications that it is possible to make with this substance? Definitely not! Whatever will be attempted by researchers with a view to new applications, I will applaud their success, just as all my sympathies will be for their work which, by popularising this substance in so many interesting respects, will encourage its adoption in industrial practice.] He patented the use of ammonia for refrigeration in 1870 but there is no indication, despite his support for ammonia in industry in many diverse applications, that he commercialised the ammonia vapour compression system at that time.

A frequently repeated myth about ammonia refrigeration is that it was the brainchild of the German scientist Carl von Linde who built a machine in 1876 for a Trieste brewery and kick-started a successful industrial gases and engineering company which is still active, with sales of over $30 billion in 2021.

Linde was certainly an early adopter of ammonia but his first attempts at mechanical refrigeration, beginning in 1873 and following Charles Tellier's lead, were with dimethyl ether as the working fluid. Mikael Hård notes in his biography of Linde, "Machines are Frozen Spirit" (Hård, 1994), that he only switched to ammonia as a safer alternative when his prototype ether system exploded and completely demolished his laboratory. Linde's great contribution was, as Hård observes, that he was an unusual combination of scientist, engineer and businessman. He was able to address the academic challenges of the new science with sufficient technical skill to turn theory to practice and with the commercial acumen to make it a profitable venture.

Some other sources, for example Woolrich (1965), Thévenot (1977) and Lindborg (2007), cite the Scottish emigrant, David Boyle, as the first to patent an ammonia compressor for refrigeration, giving his US patent number 128,448 of 1872. However patent 128,448, "Improvement in Ice-Machines", although it contains several novel features, does not specify which refrigerant is to be used in the system. To assess the relative merits of these competing claims it is helpful to understand something of the contenders' life stories and to measure their achievements according to the statements made about them by their contemporaries and in their own words.

## David Boyle

Boyle, Figure 1, was born in Johnstone, Scotland on October 31, 1837, the third of six children of David and Agness. The 1841 census lists four children (John, Mary, David and James) living with David, a grocer and his wife at an address on High Street, Johnstone. Ten years later two more children (Robert and Agnes) have been added to the family but the head of the household is described as a widower. Mary, by then aged 17, is listed as the housekeeper and the eldest son, John, is no longer resident with the family.

**Figure 1 – David Boyle, 1837 - 1891**

Fowler's Directory for Paisley and Johnstone in 1851 describes David Boyle senior as *"Grocer, Spirit Dealer and Cork Cutter"* with premises at 71 High Street, Figure 2. A John Boyle is listed in the 1851 directory as *"Butter and egg merchant"* in premises at 22 High Street.

**Figure 2 - High Street, Johnstone, home of the Boyle family, as it was at the end of the 19th Century**

The town of Johnstone, in the district of Renfrew, is unusual among Scottish towns because it was a planned community, created in the late 18th century to supply housing for workers in an industry that had not yet been established. It therefore has quite an American layout, with a main street (High Street) leading through a town square and with a grid pattern of side streets as shown in the early map from the 1850s in Figure 3. The population grew from 10 people in 1780 to 1,434 in 1792 and by 1811 it was 3,647. This planned growth was the work of George Houston, the 4th laird of Johnstone, who intended to develop cotton mills on the banks of the Black Cart river, which runs to the north-west of the town. Two mills were constructed in the early 1780s close to the Brig O' Johnstone using overshot waterwheels and the newly constructed town provided homes for many of the mill workers.

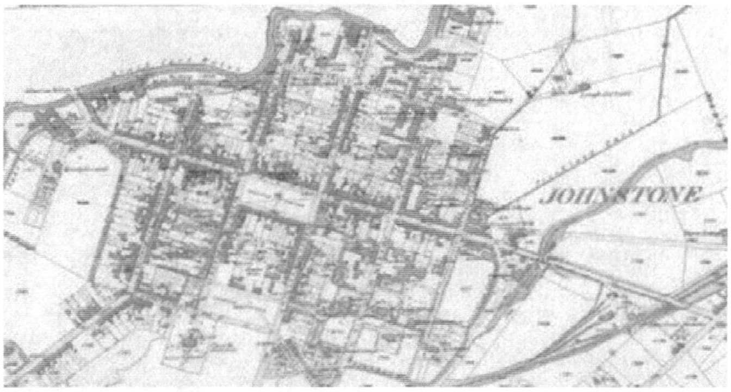


**Figure 3 – Map of Johnstone 1782 and Ordnance Survey map, surveyed in 1857, 25 inch to the mile**

The world's first machine tool factory, Craig and Donald, was opened in Johnstone in 1815, Figure 4, and the town was rapidly becoming known as a centre of excellence for machine tools, particularly related to millwork and shipbuilding. By 1837 when David Boyle was born, there were 11 mills in Johnstone, all powered by the Black Cart, producing mainly cotton and thread but also paper, leather goods and other fabrics (Scott, 1951).

**Figure 4 – Memorial to Craig and Donald in William Street, Johnstone and a 1966 picture of the site**

At the age of 15 Boyle was apprenticed as a grocer together with his brother, James, who was one year younger. It is suggested that they would both have preferred to train as mechanics but their father overruled them and so they followed in his trade. In 1859, when David was 22, he left Scotland, possibly with James, and emigrated to Mobile, Alabama, a port on the Gulf of Mexico. He set up business, hoping in time to earn enough to establish his own grocery store. At that time Mobile had a population of about 32,000 and was one of the main ports serving the south. The Civil War, which started in 1861, ravaged the southern states over the next four years and brought mixed fortunes for the Boyle family. Towards the end of the war, David had moved to Demopolis, a small town about 150 miles north of Mobile, and found that he could earn good money using imported natural ice to make lemonade. However the high cost of transporting ice from the north and uncertainty over the supply persuaded him that there would be a high demand for manufactured ice if an economic and reliable means of production could be found. This started a ten year mission to find an improved ice machine and took him and his family from Demopolis to New Orleans, then to San Francisco, then back to New Orleans, then to Jefferson, Texas (a town of about 4,000 people 150 miles east of Dallas, near the Louisiana border) and finally to Marshall, Texas (population about 2,000, 15 miles south of Jefferson). In New Orleans he learned about absorption ammonia machines but concluded that they were too expensive to build and too expensive to operate, so travelled to San Francisco seeking more information. There he ordered a Siebe machine from London, based on Harrison's ether machine and quite popular in England, but after waiting a year for it to be shipped he sued to get his money back. During his time in San Francisco he worked on better ways of making ice, resulting in his famous patent (Boyle, 1872) but his funds ran out, so the family moved back to New Orleans in November 1872 where he attempted unsuccessfully to licence the technology of his improved ice machine. In June the following year the Boyles relocated to Jefferson, with David perhaps anticipating the opportunity to build his own machine with the prospect of higher sales of ice there and he tried to get the machine to work reliably. By this stage, according to his own account of his developments (Rushton, 1896), he was working with his brother James who went on to patent an ice machine compressor in 1876 (Boyle, 1876). The work in Jefferson was painfully slow and difficult – the machine when constructed and charged with ammonia leaked "like a sieve" (Nickerson, 1891) and the summer passed without any sales of ice. They moved the prototype to Marshall, at that time a smaller town than Jefferson, but growing rapidly due to cotton trading. The ice machine finally ran reliably in April 1874 and sales picked up over the summer. In August that year a fire destroyed the ice plant and Boyle travelled to Quincy, Illinois, a city of about 25,000 people on the Missouri border, strategically located on the Mississippi river and the railroad, with the new Quincy Rail Bridge over the river providing valuable links to Chicago in the east and Kansas City in the west. Two ice plants were constructed there, funded by local businessman William Bushnell for installation in Texas. Boyle relocated temporarily to

Washington DC in October and November, returning to Quincy in the spring of 1875 before finally moving to Chicago in August 1875, where he set up in business as The Boyle Ice Machine Company. The Boyle Company and the Boyle family had finally found a home.

In the next few years, the Boyle Ice company grew steadily and by the end of the decade they had installed 7 ice machines in Texas, one in Atlanta and one in Adelaide, South Australia. They had also developed a water chiller for brewery applications, with a trial unit installed in a brewery in Chicago in July 1878.

**Figure 5 - the offices at 10 N Jefferson St, Chicago of the Boyle Ice Machine Company**

Boyle's security in Chicago was founded on the basis of having persuaded the Crane Brothers Company to act as his manufacturing base, so the Boyle Ice Machine Company and the Crane sales office shared the address at 10 N Jefferson St (now 156 N Jefferson St). Jacob Skinkle, Treasurer of Crane Bros, also acted as President of the Boyle Ice Machine Company, with William Bushnell as the company secretary. The original Crane Brothers office building is still intact (Figure 5), although the factory behind it has been replaced with apartments.

David Boyle's patent 128,448, titled *"Improvement in Ice Machines"*, which is often cited as the invention of the ammonia compressor, was filed while he was a resident of San Francisco and was granted on June 25th 1872. It is remarkable because it doesn't mention ammonia and is not concerned with the compressor used, although it does relate to ice machines, "particularly to that class in which the vaporization of a volatile fluid is used for producing the required degree of cold". Of course this does not mean that he wasn't contemplating the use of ammonia, or even already using ammonia, at that time although circumstantial evidence suggests that he didn't start those tests until he was back in Texas. The key elements that form the claims of patent 128,448 relate to the evaporator, which was configured to make plates of ice using direct expansion refrigerant fed through a combined expansion valve and distributor. The system included a method of reverse cycle defrost to release the ice plates and a glass level gauge on the outlet of the condenser to enable the manual expansion valve to be set accurately. The compressor is barely mentioned in the text and is only referred to as an *"exhaust and force pump"*. Note that the "exhaust" is what we would now call the suction of the compressor.

Boyle's other three patents were all also titled *"Improvement in Ice Machines"* and were granted consecutively on 11 May, 1875 although they were filed from various places at different times over the preceding six months. Patent number 163,142 describes a method of agitating the water to be frozen and the refrigerant to ensure clear ice is formed quickly and evenly (Boyle, 1875). It was filed from Washington DC in November 1874. Usually the two witnesses required are unknown names, presumably clerks in a legal office, but the witnesses to this invention were Boyle's financial backer and Company Secretary William Bushnell and an engineer called Thomas Rankin who went on to design many significant refrigeration systems over the following two decades, including the Ice Railway exhibit at the Chicago World's Fair in 1893 which showcased the De La Vergne compressor technology.

The other two patents granted on 11 May were 163,143 on further developments of the freezing plates, distributor and defrost mechanism and 163,144, which describes a single evaporator plate in more detail. The witnesses for these two patents were not recognised players in the ice making industry.

## James Boyle

Much less has been written about James, David's younger brother, except that he is the one who patented the compressor (Boyle, 1876). This was filed on November 24, 1875 and notes James as resident of Houston, Texas. It was granted on March 21, 1876, almost a year after David's three patents on Improvements in Ice Machines and nearly four years after the original San Francisco patent, number 128,448. James' patent, number 175,020 is titled *"Improvements in gas-liquefying pumps"* and also doesn't mention ammonia although it is suggested by various sources writing around the time of David's death that James patented the design developed by the brothers for their successful ammonia system as constructed in Jefferson and operated in Marshall (Rushton, 1896 and Nickerson, 1891). Victor Becker provides some further detail in a series of articles written for Ice and Refrigeration in 1909. He mentions several visits to Houston to inspect *"the first single-acting ammonia compressor I had heard of"* which he describes as *"a crude piece of mechanism"* but admits that it *"contained pregnant germs and it really proved to be an epoch maker, for from this primary beginning grew the remarkably successful career of the Boyle Ice Machine Co."* Becker notes that *"the inventor's brother David achieved considerable distinction and profit with it after the early death of James."*(Becker, 1909b). James Boyle made a contract with Thomas Rankin to get his patent application funded and the papers were duly submitted on November 24, 1875. However James immediately suffered a short illness and died just five days later on 29 November. Rankin continued to fund the patent in accordance with his commitment and it was duly granted in March 1876. The valuable innovation in patent 175,020 was the design of the spring loaded poppet valves for the inlet and discharge. It seems that Thomas Rankin believed he had the rights to this valve design and developed compressors with the De la Vergne Company in the 1880s, but David Boyle, who had used the design in his original ice machine back in 1873, continued with this technique in his machines from 1876 onwards. The Boyle Ice Machine Company merged with the Empire Refrigerating Company of St Louis in 1884 to form the Consolidated Ice Machine Company and when they were declared bankrupt in 1890 a dispute flared between the De La Vergne Machine Company and John Featherstone's Sons who had purchased the rights to the Consolidated Ice Machine Company's range of products. This case fizzed for the next two years, ultimately reaching the US Supreme Court, and only petered out when the patent expired on 21 March 1893.

The *"crude piece of mechanism"* is shown in Figure 6. The cylinder is mounted vertically and the valves of the compression chamber are on the base with the suction gas flowing into the upper part of the compressor before passing to the compression chamber. The connecting rod protrudes from the top of the assembly unlike most subsequent designs, which were either driven from below or mounted horizontally.

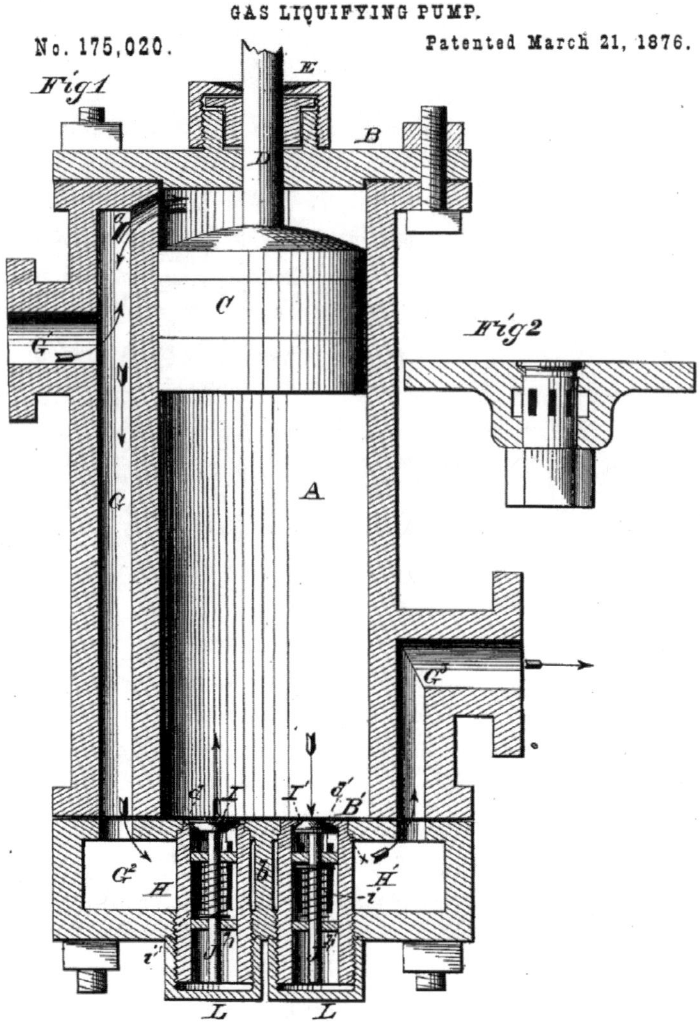


**Figure 6 – James Boyle's compressor**

The reason for the unusual suction gallery arrangement is unclear. It may have been to provide some cooling to the stuffing box on the connecting rod or to maintain it at low pressure to minimise leakage, which had been the major challenge that the Boyles had to overcome in 1872. It would have raised the temperature of the suction gas and hence reduced the effective mass flow of the compressor by reducing the suction gas density. This would not have mattered much initially, when there was no previous machine for comparison, but would have become a disadvantage as improved machines with better performance came to market. David Boyle apparently used this design for his first installation but once he was established with the Crane Brothers Company he did not persevere with it and the Boyle Ice Machine catalogue of 1879 drew the reader's attention to their *"improved apparatus for the manufacture of ice"* with an engraved plate as shown in Figure 7. This appears to show in the foreground on the left a steam engine fed with steam from the boiler at the back of the room and with an ammonia single acting, single cylinder compressor behind the steam engine. The belt and pulley arrangement drives the small condenser water pump at the back of the room and the ice is made in rectangular blocks on the right.

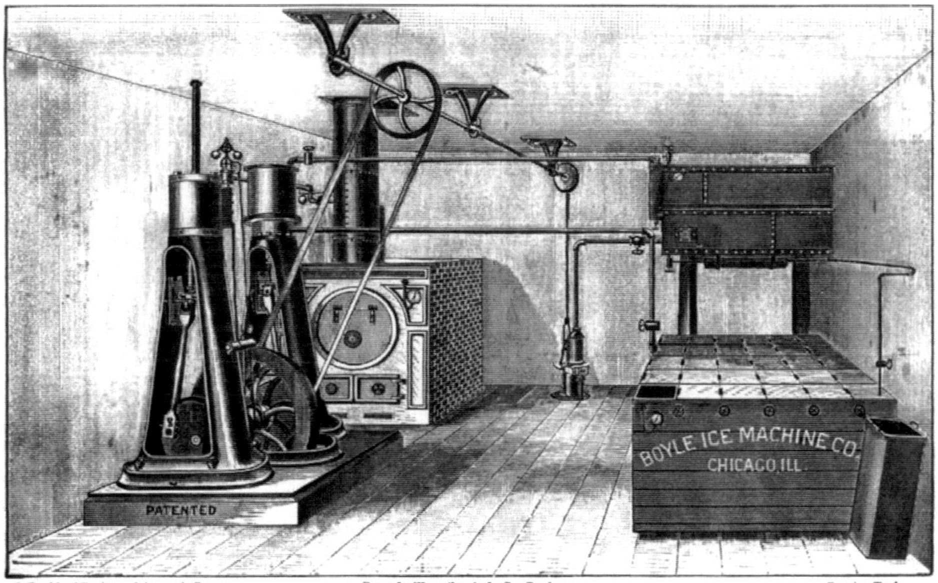


Combined Engine and Ammonia Pump. Pump for Water Supply for Gas Condenser. Freezing Tank.
ONE-TON SIZE, BOYLE ICE MACHINE.
**Figure 7 – The Boyle Ice Machine of 1879**

The Boyle Ice Machine Company is rightly regarded to be the first successful venture constructing ice making plants on a large scale. Woolrich (1965) described the company as *"the principal builder of ammonia compression machines from 1876 to 1884"* and said of David Boyle that he *"contributed more to the early machine development of commercial ammonia compressors than any other one man".* However several other innovative engineers were active in the field at the same time.

## Francis de Coppet

De Coppet was a resident of New Orleans in the 1860s and worked on early cooling experiments at the Louisiana Ice Company and the Merz brewery. Some of his experiences are detailed in an article he wrote for Ice and Refrigeration (De Coppet, 1892). He had been a draughtsman at the US Patent office in Washington DC and served in the US Army as Assistant Engineer before moving to New Orleans (Becker, 1909a). At that time the Louisiana Ice Company ran the largest ice making plant in the world at Gretna, Louisiana (now a suburb of New Orleans on the opposite bank of the Mississippi from the downtown district), equipped with six 10-ton Carré absorption machines imported from France in 1868. The success of the ice company encouraged many other local businesses to try to make their own ice, including George Merz, a brewery owner in New Orleans. He took delivery of a small compressor purchased from Charles Tellier in 1869 and supplied by Leopold Bouvier, Tellier's agent in the United States. This machine had been designed to work on methyl ether and an attempt was made, unsuccessfully, to operate it on anhydrous ammonia. The high pressure required and the relatively high speed for the time of 350 revolutions per minute resulted in frequent breakdowns so in the autumn of 1869 Merz invited De Coppet, who had been involved in the Louisiana Ice Company trials, to design an improved ammonia compressor. This first compressor was a single cylinder horizontal double acting machine and was used to chill air which was forced through tubes in a flooded shell and tube evaporator to chill a beer cellar in the brewery (De Coppet, 1892)

The compressor worked well but to maintain the beer cellar at the desired temperature of 40°F (4.5°C) the air had to be chilled to 32°F (0°C) which caused frequent blockages of the horizontal tubes due to ice build up.

De Coppet applied for a patent for his compressor in 1871, also under the title *"Improvement in Ice Machines"* and it was granted, number 148,675 on March 17, 1874. The novelty in De Coppet's design, and the focus of his claims in the patent, is the use of glycerine, or propylic alcohol as De Coppet calls it, to improve the shaft seal arrangement. De Coppet describes this in the patent text as follows:

*"A quantity of glycerine – say several gallons at one operation – is put in the cup S; then screw on the cover of the cup;then open the small cock between the cup and the induction pipe G; the glycerine will then flow by gravity through the receiving side pipe and valves in the pump, and the movement of the pump-piston forces the glycerine around the piston-packing, filling the receptacles or grooves in the packing rings and the valve and their seats with glycerine, and any surplus glycerine is forced on through the delivery side pipe, in and through pipe H in trap E, where it falls by gravity to the bottom of the trap, the ammoniacal vapor continuing on to the compression-coil for compression to liquefaction."*

This is illustrated in Figure 8, taken from Patent 148,675

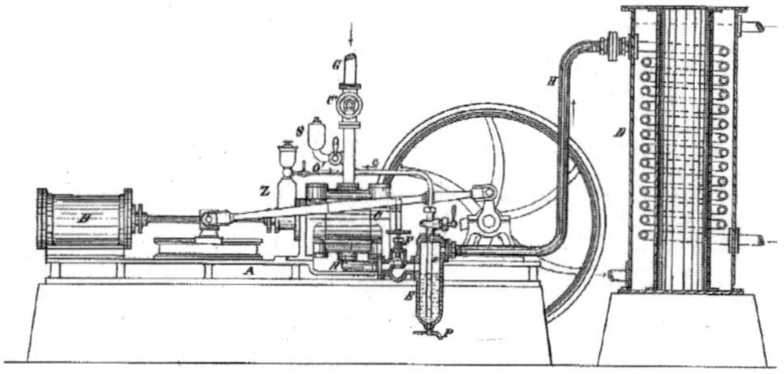

**Figure 8 – De Coppet's glycerine sealed ammonia compressor from his 1871 patent application**

This is a significantly more sophisticated device than James Boyle's *"crude piece of mechanism"* although it predates Boyle by at least a couple of years. Becker (1909a) noted that De Coppet, having worked in the Patent Office as a draughtsman, *"was a much better draughtsman than he was a specification writer"* and goes on to say *"the drawings of his patent therefore better illustrate his invention than does the description embodied in the specification; together they describe the invention in accordance with the rules of practice of the U.S. Patent Office then in force, but through having been indifferently prepared the claims did not give him all he was entitled to at the time"*. This seems a rather harsh assessment. De Coppet followed the "problem-solution" approach to patent claims where a problem with the current state of the art is described, a solution is explained (or "taught" as the patent lawyers would say) and the details of the solution are then claimed. The claims are the only part of the full patent text that provides the legal status and a common mistake is to claim too much of previous devices, leading to rejection of the application on grounds of insufficient novelty. De Coppet neatly avoided this pitfall by concentrating on the problems of sealing leaks on the piston rod stuffing box.

It can be concluded that this was probably a large part of the problem encountered at the Merz Brewery with the Tellier machine operating on ammonia and although Becker describes the De Coppet compressor as *"the first ammonia compressor ever built"* (Becker, 1909a) this is clearly not the case, as the Tellier machine on trial at the Merz Brewery in 1869 predated it. It could however perhaps be described as "the first purpose-built ammonia compressor".

De Coppet licenced manufacturing rights to his patent to the Fred Wolf Company of Chicago and the De La Vergne Company of New York. Wolf later adopted the Linde compressor design and manufactured compressors in Chicago on behalf of Linde from 1881 onwards. De La Vergne built a trial machine on the glycerine sealed principle in 1878 for his New York brewery and spent the next four years refining the design before entering the market as a compressor manufacturer in 1882. Over the next eight years the De La Vergne company delivered over 250 compressors and established sub-licenses in several countries, including L Sterne and Co in the UK. De La Vergne machines were installed in that first 8 year spell in England, Scotland, Canada, Argentina, Ecuador, Brazil, Italy and the Philippines as well as all across the United States, all with De Coppet's glycerine system although it is not given credit in the De La Vergne company literature (De La Vergne, 1891).

## John Beath

Beath was a millwright who worked on machinery in gold and silver mines in California and Nevada. He realised in the late 1860s that the high price of ice in San Francisco and Los Angeles, up to 5 cents per pound, was an opportunity to develop an ice manufacturing industry on the West Coast. With a financial partner, Samuel Martin, he developed and patented a novel form of evaporator, relying on the two-phase flow of volatile refrigerant through the inside of horizontal tubes. Patent no 127,180 was granted on May 28, 1872 (Martin and Beath, 1872) and suggests that vapour compression was the means of cooling the refrigerant tubes. The patent states *"At present we prefer this fluid to be ammonia, which we obtain from the aqua ammonia of commerce by distillation, usually drying the gas as it goes over by passing it through quicklime. We introduce it to our machine by connecting a pipe with our distributor, allowing the gas to drive the air before it and out through a cock in the receiver".*

Beath erected a small machine in Los Angeles in 1869 and then, with Martin, applied for the patent mentioned above on the design of the evaporator. According to Beath's account (Beath, 1912) the pair applied for a second patent but then, when they moved to San Francisco to build a larger machine, Martin announced that he had no more funds due to recent losses in his mining operations. Beath managed to find other backers for the San Francisco project and it was successfully completed. Hearing of high ice prices in Portland, Oregon, Beath then moved there to construct another larger ice machine. However shortly before completion the plant was destroyed in a large fire. This might have been the devastating blaze of August 1873, which razed 22 city blocks in the centre of Portland, but was probably an earlier blaze in December 1872 or even earlier. Beath had been able to study the behaviour of the evaporating ammonia in the tubes and constructed a second machine on the site of the first but with further improvements in the evaporator design. He then travelled to Chattanooga, Tennessee and built an ice plant based on absorption but with his new evaporator design before returning to San Francisco to upgrade the plant there with his improved system. In 1874 he was drawn into a patent dispute with his former partner Samuel Martin. Martin was successful although Beath maintained that the claims of Martin's patent were stolen, and Beath noted of Martin that this *"was the turning point in his career; he lost his money, took to heavy drinking and died in an insane asylum"* (Beath, 1912). Despite his early experiments with ammonia compressors, all of Beath's successes seemed to be with absorption systems and following his Chattanooga installation he noted that *"my plan is now largely used all over the world".*

Two other eminent engineers are worth mentioning at this point. Alexander Twining, a civil engineer from New Haven, Connecticut worked on artificial ice manufacture from about 1850 onwards and was granted a patent, number 10,221 in 1853 on a vapour compression system using *"a volatile liquid, as alcohol, ether, sulfuret of carbon etc, of which at present I find ether the most available"*. Twining constructed a machine in Cleveland Ohio using diethyl ether and it ran successfully for many years. He suggested several improvements over the next few years, including the addition of an ammonia compressor to a Carré absorption system, Patent number 121, 975 dated December 9, 1871. The French engineer, Charles Tellier is credited with being the first to equip a ship with refrigeration equipment for the transportation of meat over long distances. Tellier favoured ether for this purpose, and according to Thévenot (1977) installed his first ether compressor in a chocolate factory near Paris in 1868 and attempted a shipment of meat from the Americas but this was unsuccessful. It was 1876 before he made a successful shipment from Buenos Aires to Rouen. From the beginning of his investigation of refrigeration systems Tellier was interested in the opportunities offered by ammonia in vapour compression although his early machines all used ether. At a paper presented to the Society of Arts in London in 1868 the speaker, Dr B Paul, identified three methods of artificial freezing; vapour compression of a volatile liquid, absorption of ammonia and compression and expansion of air (Paul, 1868). In the paper the only volatile fluid mentioned in connection with vapour compression was ether but the discussion afterwards centred largely on the various schemes being tried by Charles Tellier, including his investigation into the storage of meat. In this discussion a Mr Shand said that he had visited Tellier and seen his equipment in operation, using ammonia as the refrigerant. This statement was later repeated by the speaker who stated that Tellier *"used anhydrous liquid ammonia in the same way as ether was used for the reduction of the temperature by evaporation and then re-condensed the ammonia by means of mechanical power"*. Tellier was a well known fan of ammonia. His major work, Ammoniaque Dans L'Industrie was published in 1867 (Tellier, 1867) and included a vast array of possible uses for this remarkable substance, including as a safer alternative to hydrogen in lighter than air balloons, as a fuel for motor vehicles, as a refrigerant in absorption systems and as a refrigerant in what he called *"mechanical compression apparatus"*. The applications for refrigeration that he described include cooling of air in theatres, desalination of sea water and the manufacture of copies of marble statues as well as ice making, food preservation by chilling in warehouses and comfort cooling for passengers at sea.

Despite the views of the London audience on Tellier's use of ammonia in his compressors it seems that his machines were not well adapted for this purpose. When the installation at the Merz Brewery in New Orleans was modified to run on ammonia it was a singular failure, suffering regular breakdown of the connecting rod owing to the high forces involved in compressing ammonia compared to dimethyl ether and apparently also leaking profusely from the shaft seal. All of Tellier's marine installations used dimethyl ether and his original patent didn't mention ammonia. The received wisdom at the time was that ammonia was an excellent refrigerant but could only be used in Carré's absorption system and when the trial at Merz suggested that compression of ammonia was feasible, Leopold Bouvier quickly persuaded Tellier to apply for patent protection for this development. US Patent number 100,689 was issued on March 8, 1870 claiming

1. *"The use or application, for the purpose of generating artificial cold, of pure ammoniacal gas liquefied by means of mechanical compression, substantially as described,*
2. *The use or application, for the purpose of liquefying ammoniacal gas, of the pump and condenser, forming part of the machines or apparatus fully described in Letters Patent No 85,719 issued to me January 5, 1869, substantially in the manner and for the purpose above set forth."*

In the teaching of Patent 100,689 Tellier wrote *"Liquefied ammonia, prepared and used as herein described, possesses great advantages over the various substances heretofore employed for cooling or refrigerating."* These substances, according to Tellier, were sulphuric ether, hydrocarbons and carbonic acid.

## Discussion

Assessing the achievements of engineers of so long ago is difficult. Some clues are given by what they wrote about themselves, but these accounts always only give one side of a story, particularly over a point of contention. Other hints are found in the patents that were granted to them, but it must be remembered that not everything that a person does is claimed in a patent and not everything claimed in a patent has actually been done in practice. It is interesting to note that the compressors described in the various patents studied here seem to fall into two sorts – practical machines, like those of Alexander Twining, Charles Tellier, Francis De Coppet and James Boyle and gross oversimplifications like the "exhaust pumps" described by Beath, Martin and even that shown in David Boyle's patent 128,448. A more accurate picture can be gained from the things that colleagues and acquaintances wrote about these pioneers although this too can be patchy and misleading. Some mistakes and half truths get picked up and repeated over and over and some people are forgotten and slip out of the narrative over time.

David Boyle was described in an 1885 article in The Western Brewer as *"a man of sterling integrity, quick, sensitive and eager to succeed. His every thought is towards the advancement of the world's knowledge of the science of artificial refrigeration."* His obituary in Ice and Refrigeration in 1891 said *"few have obtained as thorough a knowledge of the science of artificial ice making and refrigeration"* and a reminiscence written in 1896 described him as *"the father of successful artificial refrigeration and ice making: for although many had spent years of toil and vast sums of money in experiments in the art before he entered the field, yet to David Boyle is due the honour of having accomplished more in the advancement of the science than the combined results of the efforts of all of his predecessors".*

Others were described in less flattering terms. John Beath noted that Samuel Martin, his former business partner died in an insane asylum and Victor Becker described Leopold Bouvier as "the Casanova of mechanical refrigeration", not intended as a compliment. Becker also noted that Bouvier *"decamped with the proceeds of a draft for the last payment"* on a project in Newark in 1877 and *"after several other adventures finally hung himself with his suspenders in a prison cell".*

## Conclusions

A number of notable "firsts" have been identified in this study. All of them are prefixed here with the word "probably".

Alexander Twining was probably the first person to use vapour compression to manufacture ice in North America.

Charles Tellier was probably the first person, together with his associates to run an ammonia compressor for this purpose. This may have been around 1867 in Paris or perhaps around 1869 in New Orleans.

Charles Tellier was probably the first person to patent an ammonia compressor with the granting of Patent number 100,689 on March 8, 1870.

Francis de Coppet was probably the first person to build a reliable, leak-tight, robust ammonia compressor with his glycerine-based adaptation of the Tellier machine as shown in Patent number 148,675, granted on March 17, 1874.

James Boyle was probably the inventor of the spring loaded compressor valve, as shown in Patent number 175,020 of March 1876

David Boyle was probably the first person to make the construction of ice machines based on the compression of ammonia vapour a commercial success.

## Acknowledgements

Bernard Nagengast, co-contributor for this paper, who has been interested in refrigeration history for 55 years, provided much of the basic research that led to the conclusions offered here. Andy Pearson sifted through that material and added some observations and insights. The authors would like to thank Iain Murray and his colleagues at the Johnstone History Museum for a wealth of information about 19th Century Johnstone and the directors of Star Refrigeration Ltd and ASHRAE for their ongoing support. Both of the authors are very grateful to Tracy Boyle Abrusci, great-great-granddaughter of David Boyle, and John Boyle, great-great-grandson of James Boyle for their additional information and insights, which helped to clear up several long-standing and persistent mysteries.

## References

Beath, J., 1912, *"Ice making experiences",* Ice and Refrigeration, September 1912, p81

Becker, V., 1909a, *"Elemental Processes Related to Mechanical Refrigeration"*, Ice and Refrigeration, March 1909, p108

Becker, V., 1909b, *"Elemental Processes Related to Mechanical Refrigeration",* Ice and Refrigeration, May 1909, p252

Boyle, D., 1872, *"Improvement in Ice Machines",* US Patent No 128,448

Boyle, D., 1875, *"Improvement in Ice Machines"*, US Patent No 163,142

Boyle Ice Machine Co, 1879, *"Boyle Ice Machine and Refrigerating Apparatus"*, Catalogue published December 25th 1879

Boyle, J., 1876, *"Improvement in Gas Liquefying Pumps"*, US Patent No 175,020

Calm, J., 2012, *"Refrigerant Transitions ... Again",* Proc ASHRAE/NIST Refrigerants Conference, Gaithersburg

De Coppet, F., 1874 *"Improvement in Ice Machines"*, US Patent No 148675

De Coppet, F., 1892, *"Pioneer Ice Making"*, Ice and Refrigeration, January 1892, p37

De La Vergne Refrigerating Machine Co, 1891 *"Mechanical Refrigeration Processes and Apparatus"* Catalogue published January 1890, New York

Fowler, G., 1851, *"Paisley and Johnstone Commercial Directory and General Advertiser for 1851-52"*

Hård, M., 1994, *"Machines are Frozen Spirit"*, Campus Verlag, Frankfurt, p134

Lindborg, A., 2007, *"Ammonia's Future as a Refrigerant"*, Proc IIR Conf Ammonia Refrigeration Technology for Today and Tomorrow, Ohrid

Martin, S. and Beath, J., 1872, *"Improvement in the Manufacture of Ice and Refrigerating Machine"*, US Patent No 127,180

Nickerson, J., 1885, "David Boyle, Esq., Chicago", The Western Brewer, October 1885, p1731

Nickerson, J., 1891, "The Late David Boyle", Ice and Refrigeration, July 1891, p24

Paul, B., 1868, "Artificial Freezing and Refrigeration", Journal of the Society of Arts, p67

Rushton, W., 1896, "A Reminiscence", Ice and Refrigeration, March 1896, p158

Scott, C., 1951, "Burgh of Johnstone Souvenir Brochure Festival of Britain" Burgh of Johnstone Town Council

Tellier, C., 1867, "Ammoniaque dans L'Industrie", Paris

Tellier, C., 1869, "Improvement in the Manufacture of Ice and the Refrigerating of Air, Liquids etc", US Patent No 85,719

Tellier, C., 1870, "Improvement in the Manufacture of Ice and Cooling Air, Liquids etc", US Patent No 100,689

Twining, A., 1853, "Manufacturing Ice", US Patent No 10,221

Twining, A., 1871, "Improvement in Ice Machines using Ammonia" US Patent No 121,975

Thévenot, R., 1977, "A history of refrigeration throughout the world", IIR, Paris

Woolrich, W., 1965, "Chicago-Milwaukee Pioneers in Mechanical Refrigeration Design 1867 - 1887", Address to the ASHRAE Life Members Club, Chicago, held in ASHRAE Library and Historical Archive, Atlanta

## Picture Sources and Credits

Figure 1: Nickerson, J., 1885

Figure 2: Scott, C., 1951

Figure 3a: Scott, C., 1951

Figure 3b: National Library of Scotland

Figure 4a: © Campbell Crombie

Figure 4b: Historic Environment Scotland © HES. Reproduced courtesy of J R Hume

Figure 5: © Google Streetview

Figure 6: Boyle, J., 1876

Figure 7: Boyle Ice Machine Co, 1879

Figure 8: De Coppet, F., 1874

## About the authors

### Dr Andy Pearson FInstR

Group Managing Director of Star Refrigeration is a past president of the Institute of Refrigeration and past-chairman of the Institute of Refrigeration's Technical Committee. He has served two terms as a director of the International Institute of Ammonia Refrigeration and is a member of the Refrigeration Safety Technical Committees for the British Standards Institute (BSI), the European Committee for Standardisation (CEN) and the International Standards Organisation (ISO).

### Bernard A. Nagengast

Bernard Nagengast is a consulting engineer and the winner of the 2020 Engineer-Historian Award presented by the American Society of Mechanical Engineers and is a co-author of the Book "Adventures in Heat and Cold: Men and Women who Made Your lives Better". He is also an ASHRAE Distinguished Life Member.

## DISCUSSION REPORT

**Dermot Cotter**, chairing, thanked the author for a very interesting paper.
My first question is about David Boyle, he was born in Scotland, did he ever come back to Scotland in his lifetime?
Andy Pearson – As far as I'm aware when he left in 1859, he never came back. That was very common at that time. He left at the age of 22, never to return.

**Dermot Cotter** – So all his work really shows up in the US?
Andy Pearson – There is a reason why the American TV and movie industry always casts a Scot as the engineer and David Boyle is probably part of that story.

**Dermot Cotter** – Was there anything related coming out of the UK at that time? Around David Boyle's time?
**Andy Pearson** – I focussed on the American patents because that was where Boyle was working but the first vapour compression machine was built (by an American) in London in 1834. That was Jacob Perkins using Diethyl Ether and James Harrison, in Australia developed a Diethyl Ether machine but then came to London to get it manufactured. James Harrison was also Scottish, he was born just north of the river Clyde. James Harrison came to London to get his machines manufactured and it was one of those machines that was being made by the CB Brothers that David Boyle tried to buy when he was in San Francisco. There was a lot happening in the UK and you have to bear in mind that while villages and towns in the USA had populations of 1000s, London at that time had a population of 4 million and was the biggest city in the world and was the absolute hub. There was a huge amount happening and Germany and France were leading innovation as well. The need for ice in the USA was far greater so ice manufacturing was what sparked the development and then other industries like brewing and meat packing and transportation saw the possibilities and they latched on to the new technology of refrigeration.

**Dermot Cotter** – These days thousands and millions is spent on R & D. Do you see the early pioneers just taking a chance to see what happens?
**Andy Pearson** - That's what I was hinting at when I said that I would characterise that first generation as 'whatever didn't work'. Things that these people achieved on the basis of no information at all was quite incredible but Carl Linde stood out from the rest because he was a fantastic scientist first and foremost with a good engineering knowledge. Other people like Lord Kelvin, a fantastic scientist but I would say his natural engineering was perhaps a bit theoretical and he got involved in lots of business ventures but it was always other people's business ventures. Lundy had that unique combination of scientist, engineer and businessman and that was why he was so successful but that was unusual and the majority of people were either scientists trying to make refrigeration work or engineers working with their hands and experimenting and trying to make it work. It was a difficult time for them.

**Dermot Cotter** – When you call it industrial refrigeration, what kind of Kilowatt capacity are we talking about? I think in terms of Megawatts today.
**Andy Pearson** – It grew incredibly rapidly once they had Ammonia as the working fluid. The early machines that were making 1 tonne of ice a day were using a handful of Kilowatts, maybe 10 Kilowatts of cooling capacity in total but very quickly they were making machines that would do 200/500 tonnes a day per compressor. A 500 TR compressor would be about 150 Kilowatts. These machines were the size of a bus. They were very slow running and they were powered by steam engines and the steam engine, typically, was included in the frame, along with the ammonia cylinders. There were platforms that you could climb on top of, whereas now a 150 Kilowatt compressor could be picked up with your hands now.

**Dermot Cotter** – I assume all this stuff was from steam engines and adapting components
**Andy Pearson** – A lot of the early manufacturers were farm machinery manufacturers, traction engines and that kind of thing. That's where the skills lay, to adapt them to make compressors.

**Dermot Cotter** – A question from David Blackhurst. You talked about the brilliant innovation between the 1850s and the 1870s. What about between now and the 2040s?
**Andy Pearson** – I've said for a long time, and I'm not sure I will ever see this but, hermetic oil-free compressors. We have them for fluorocarbons but, so far, not for ammonia or for $CO_2$. Oil-free is a real game changer. There are lots of reasons for having oil but one by one you can get rid of them and oil becomes less of a benefit. I think that would be something that I would really like to see.

**Dermot Cotter** – Do you have any information on the Einstein refrigerator?
**Andy Pearson** –It was developed in the early 20th century by Leo Slizzard. Einstein worked with Slizzard and it was basically a carry-style absorption system but the innovation that they introduced was to add some hydrogen to the mix, which turned the absorption machine from being something that was periodic to a continual operating machine. A great invention. I think calling it the Einstein refrigerator is latching on to a famous name because he wasn't really the key driving force.

**Dermot Cotter** – Loretta Birch clarifies that David died in St Louis and she highlights that he was in San Francisco in 1869 not 1868 but what she really is highlighting is the massive distances that he was travelling back then. Nowadays we would just get a flight but back then getting a train.....
**Andy Pearson** – I would love to know how they travelled because I think he took his family with him. He was there for about 3 years and then travelled back and this was harsh territory that they were travelling through and not at all easy. Possibly they went by rail or by wagon. This is the wild west we are talking about. We fail to appreciate that. Loretta is one of David's descendants so I'm grateful that she has joined this talk. My understanding is that David died in Mobil, Alabama as was written in the obituary. He was on a business trip and had a heart attack in 1891. He died young, just in his early 50s but was remarkably well-liked by people.

**Dermot Cotter** – Is David or James Boyle connected in any way to Boyle's Law?
**Andy Pearson** – No, there's no relation as far as I understand. Boyle's Law dated from the 1600s.

**Dermot Cotter** – When we go back to the 1800s, and I know we have talked about David Boyle, but in that generation who would be your engineering hero?
**Andy Pearson** – That's a really tough question because I admire so many of them for what they did based on very little. Some of them were quite wild characters as well. I've always had a soft spot for a man called Thadeus Lowe, who was the first person to run a carbon dioxide compressor. That was 1867 so it is exactly the time when all this stuff was going on but Lowe sounded like he had a really quick temper. He took part in a competition to ship beef into Texas using refrigeration on a ship and he lost because the boat that he picked, by the time he had put all his equipment on it, was sitting too deep in the water and couldn't get into the harbour. It got to the entrance to the harbour first and then couldn't get in to dock. He went off in a huff to Los Angeles and built an observatory on top of a mountain and went and looked at stars for the rest of his life. He ran away from a farm and the circus was the extent of his scientific training. A great development that failed to turn into an innovation. He certainly laid the groundwork for a lot of technology. He took a compressor that was used to compress Hydrogen to 1 bar absolute and took the compression to 1.1 bar absolute and used it to compress carbon dioxide from 30 bar to 60 bar. There's quite an amazing story in that from a technical point of view. In the UK, from a science point of view William Rankine, I feel, is overlooked. He worked at the same time as Lord Kelvin in Glasgow University. Both of them worked on thermodynamics and the absolute temperature scale in SI units is the Kelvin scale but the

absolute scale in imperial measure (Fahrenheit) is the Rankine scale. The two scientists each got a temperature scale named after them which is quite neat.

**Dermott Cotter** – Thank you again Andy for your insights.
**Andy Pearson** – I'd like to thank Bernard Nagengast for his help. He really set me off on this path and has been a great support all the way through.

IOR.org.uk
REFRIGERATION AIR CONDITIONING HEAT PUMPS

The knowledge hub for refrigeration, air conditioning and heat pumps

# The IOR's part in the Net Zero agenda for the built environment

**Address by Graeme Fox FInstR CEng**
**Presidential Address, 7 December 2022**

Copyright © 2022
Institute of Refrigeration

## Why you should attend

- To understand how heat pumps, heat recovery and heat networks will fit into the net zero agenda
- To take advantage of the opportunities in deploying these technologies and understand the challenges involved
- To consider more collaborative working across sectors and new approaches to skills development

## Abstract

The UK Government has set a target of achieving net zero carbon emissions by 2050 as part of their attempt to address the climate change emergency and the associated impact. The built environment is responsible for around 40% of total carbon emissions, and an environmental audit committee report highlighted that 3% of UK greenhouse gas emissions are of HFCs.

The Committee on Climate Change has said that to meet the commitment to reach net zero by 2050, 19 million heat pumps will need to be installed and that hybrid heat pumps should be widely used by 2035.

Heat pumps are not a new technology – they have been used in a variety of different applications for many years. However, their increased use comes at a time when the legislative landscape is looking increasingly at restricting or limiting the use of refrigerants that have a direct environmental impact. This places additional considerations on the building services designer with regard to which refrigerants can be used and what the likelihood is of restrictions on their use over time.

Heat recovery is also not a new concept – it has been used very effectively for decades in a variety of applications and incorporating different technologies to achieve and deliver low carbon solutions before we even thought that "low carbon" was necessary. We did it because it saved money.

Our sector – representing industrial and commercial refrigeration as well as building services air conditioning and heat pump applications – needs to approach projects with a far more collaborative attitude going forward if tomorrow's buildings are to be truly carbon neutral. Megawatts of usable heat is being rejected from cooling applications every year and much of this could, with a little bit of joined up thinking, be recovered and used to supply heat for heat networks or social housing built near the non-domestic buildings. The building services design consultants need to work more collaboratively with the refrigeration consultants, rather than in the traditional silos, so that we can harness rejected heat and help our built environment become truly carbon neutral.

My paper will discuss these options and offer some solutions by example and with reference to case studies which have proven that the technology and applications work in the real world.

## The Context for Net Zero

The UK Government has set a target of achieving net zero carbon emissions by 2050 as part of their attempt to address the climate change emergency and the associated impact. The built environment is responsible for around 40% of total carbon emissions, and an Environmental Audit Committee report[1] highlighted that 3% of UK greenhouse gas emissions are from HFCs. An updated report published 1st February 2022 claimed that the total GHG emissions had dropped by 9.5% compared to the 2019 figures[2] and highlighted the overwhelming bulk of these emissions is still carbon dioxide – accounting for around 79% of the total emissions.

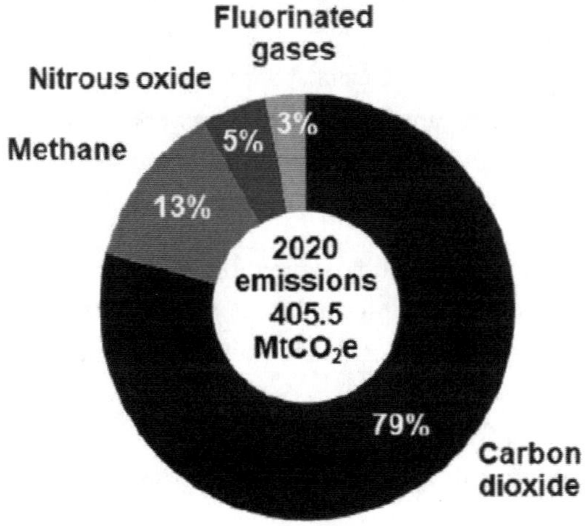


Figure 1: UK territorial greenhouse gas emissions, 1990-2020

Emissions from fuel combustion and product use in industrial and commercial sectors as well as F-Gas emissions from refrigeration and air conditioning in all sectors come under the general category "Business" for the purposes of recording emissions by sector. The Business sector accounted for 18% of the total UK greenhouse gas emissions in 2020 according to that same report. So whichever way you look at this, it is clear that our RACHP sector will play a major part in the strategy of minimising carbon consumption and emissions in forthcoming years and ultimately reducing our carbon use to net zero by 2050.

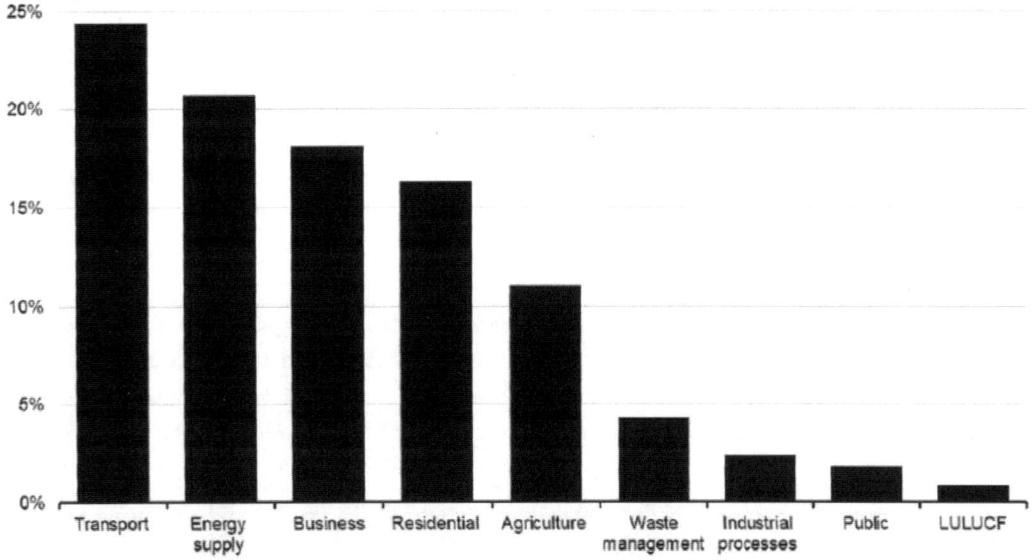


Source: Table 1.2, Final UK greenhouse gas emissions national statistics 1990-2020 Excel data tables
Note: LULUCF is land use, land use change and forestry.

Figure 2: Net territorial UK GHG emissions by sector, 2020 (%)

Total electrical consumption in the UK was approximately 301.7 TWh over the year in 2015[3] – the latest data set found – approximately 300 billion kWh of electrical power consumed. Building services alone are estimated to account for approximately 38% of this total, representing 114 billion kW of energy consumed every year.

The importance of effective planned preventative maintenance regimes is essential if the energy efficiency of a designed and installed system is to retain anywhere near the design conditions over time. This requires buy-in from clients to ensure they engage competent contractors to carry out the maintenance works, it requires maintenance contractors to invest in competent engineers and technicians in performing that work on a day-to-day basis and not corner cut at any given opportunity, and it requires equipment which performs as set out in the specifications.

The energy efficiency across the entire range of cooling plant varies enormously due to the vast range of products and applications at our disposal when we look at the whole RACHP sector incorporating industrial, commercial, retail and domestic markets. It will frequently depend on siting of equipment, air flows around heat exchangers, ambient conditions, pipe routing, etc. The actual energy efficiency performance may also be difficult to pinpoint in the real world as the system may be integrated into other plant for process cooling or heat recovery purposes, rendering the whole project more energy efficient than it may be possible to record or report.

Modern split and multi-split air conditioning systems used widely in the office and retail sectors, and increasingly in the domestic market, are all also reverse cycle heat pumps and many of them have the ability to recover heat being rejected from one zone to supply domestic hot water or space heating in other parts of the building. The standardised building energy modelling used by architects and planners has, in recent years, fallen behind the times as the product characteristic database used for SAP (Standard Assessment Procedure for Energy Performance Certification under the Building Regulations) calculations has not been able to keep apace with evolving technology. Clever bespoke designs using standard equipment in novel ways cannot be properly accounted for in the methodology, resulting in worse calculation results than can be realised in the field. This has certainly hindered progress in design concepts being more widely adopted and Governments have been slow to get to grips with how quickly our sector has evolved and delivered sustainable solutions to their clients' needs.

A lack of standards in some specialist sub-sectors has allowed disingenuous claims to be made about how effective some equipment is in reality. This is now being addressed in the heat networks sector by the creation of a Heat Networks Assurance Scheme which follows on from the Heat Interface Units (HIU) test standard developed around seven years ago and recently upgraded to improve the test process and ensure a level playing field for the HIU manufacturers to display their credentials. The ability for the BESA (the Building Engineering Services Association) HIU test results to now be directly uploaded onto the Product Characteristic Database operated by BRE on behalf of BEIS (the Department for Business, Energy and Industrial Strategy) will start to aid designers in applying the best products for their projects and ensure SAP calculations can be better made.

## Heat Pumps

The Committee on Climate Change has said that to meet the commitment to reach net zero by 2050, 19 million heat pumps will need to be installed and that hybrid heat pumps should be widely used by 2035[4].

Heat pumps are not a new technology – they have been used in a variety of different applications for many years. However, their increased use comes at a time when the legislative landscape is looking increasingly at restricting or limiting the use of refrigerants which have a direct environmental impact. This places additional considerations on the building services designer with regard to which refrigerants can be used and what the likelihood is of restrictions on their use over time.[5]

The increased use of much lower global warming potential (GWP) refrigerants across a broader spectrum of heat pump and air conditioning units is helping to maintain a healthy optimism for the continued growth of that sector – an essential element of the proposed 750,000 domestic heat pumps needing to be installed annually if we are to meet that 19 million by 2035 target.

The real issue with this anticipated uptake in the volume of work, however, is that the sector lacks spare capacity with most installer contractors reporting busy order books. A large number of traditional RACHP contractors also have no desire to work in the domestic market, where the demands on the contractor are very different to the commercial world which they are used to working in. The upskilling of traditional heating engineers who do want this domestic work is, therefore, essential going forward and huge steps forward have been made in this arena in the last two years with both BESA and the Heat Pump Association developing and delivering upskilling short courses to ensure heating engineers understand the specific design and commissioning requirements to make heat pumps work efficiently and that they understand the importance of educating the end user on the most efficient ways of running the systems. The traditional domestic heating modes of short warm up times have to be unlearned and the benefits of running at low flow temperature for longer periods of time to effect the saving of energy and money in the long run have to be understood. These will mainly be monoblock type units to allow non-RACHP qualified installers to deliver the market demands.

The next hurdle to overcome in this market will be the on-site repairs to monoblock systems. If the heating engineer has been upskilled to understand the complexities of a vapour compression heating circuit, that will not qualify or empower them to carry out compressor changes or repair refrigerant circuit leaks or faults, so whichever way you look at this market there will be an inevitable rise in demand for RACHP skilled technicians and engineers to work in the domestic market over time, and that will require a large uptake of the upskilling or cross-skilling short courses available as well as a willingness of our more traditional RAC sector to work in the domestic sector from time to time.

But it isn't just in the domestic environment where we need to see the roll out in the use of heat pumps. Commercial applications are increasingly using heat pump technology to deliver real energy savings in some retail and commercial buildings. If a large supermarket with walk-in cold stores also has staff rooms and offices in the building, why wouldn't you be looking at recovering the rejected heat from those cold rooms and use tried and tested technology to transfer that low grade heat into useful space heating rather than using more energy to create heat for this use? If you have high volumes of domestic hot water for wash down areas, why wouldn't you be looking to recover waste heat energy where you can, to raise that hot water storage temperature, minimising any additional energy you need to uplift the temperature to a safe and useful level?

## Heat Recovery

Heat recovery is also not a new concept – it has been used very effectively for decades in a variety of applications and incorporating different technologies to achieve and deliver low carbon solutions before we even thought that "low carbon" was necessary. We did it because it saved energy and, therefore, money during the energy crisis in the 1970s, and the current energy crisis is certainly starting to focus minds back onto this factor.

We have a network of readily available low-grade sources for heat pumps to tap into throughout the UK – either boreholes that have been drilled but are unused, or in the many disused coal mines. Although care must be taken with the water quality in these cases.

I was called out to a system around 10 years ago in Fife where an old, abandoned coal mine had been flooded and the water was pumped up into a heat pump which delivered heating and hot water to two blocks of social housing flats delivered as a small heat network. A great design of system, way ahead of its time, but clearly the design team had no concept of the water quality and the effect on a heat exchanger coil. Iron Sulphide reacts with air and water to create sulphuric acid with dissolved iron, and this had gradually eaten away at the heat exchanger from the inside out until the refrigerant had leaked out. We went to replace the condenser coil but then found the manifold head where the sump pump lifted the mine water to supply the heat pump was leaking and the ground around the manifold head was a rusty brown colour. A simple water test proved the solution to be highly corrosive acid which had eaten through the flange bolts so the only thing holding the flange faces together was the pipe supports around it. A sand filter and water treatment plant installation later to remove dissolved iron solids and reduce the pH of the solution and the system was ready to be put back into action.

Simple refrigeration systems are in existence in every town and city, in all manner of buildings. Butchers' cold rooms, fishmongers, convenience stores, funeral parlours, cafes, restaurants, fast food outlets, etc all have fridge and/or freezer uses, and most of that heat energy being removed as a function of the cooling process is rejected to atmosphere, even though those same properties will almost certainly have heating systems incorporated in some way, either for space and/or domestic water heating. At a time when the focus in so many of our workstreams is on minimising or at least reducing carbon consumption, it seems profligate beyond reason or measure to throw away useful heat which is a by-product of the cooling process so necessary for modern life.

Turning to the industrial and commercial sectors this wasted heat becomes even more stark. Supermarkets and food production facilities have the potential to recover megawatts of heat and turn it into useful space heating or domestic water production for common or staff areas. A lot of work is being carried out in this vein now, mainly borne out of a few large producers acutely aware of their consumers or shareholders having an eye on their organisation's Environmental and Social Governance (ESG) policies, but we could and should be going further. Feeding into heat networks has the potential to provide virtually free heating for social housing or nursing homes, for example. Our industry could thereby be providing real tangible social benefit to the most vulnerable in our society.

There is approximately 40,860,853m$^3$[6] of UK cold storage space which equates to approximately 2 million kWh of heat rejection. In simple terms this equates to the heating and hot water requirements for approximately 200,000 homes if appropriate thermal storage is used to harness the heat when rejected so it can be made available at the right time to meet demand. At a time when we are bombarded daily with media reports of the cost of living crisis, the rising energy costs crisis, and the very real prospect of homeowners not being able to afford to heat their homes to the same comfort levels we have become accustomed to, it is clear that the refrigeration sector can offer so much more to the overarching UK climate crisis net zero targets than we currently do. With a little bit of joined up thinking and collaboration between sectors more used to working in silos we can clearly help deliver affordable heating to the most vulnerable in society through social housing heat and hot water delivery.

## Heat Networks

A well-established sector outside of the UK is district heating networks. But we struggle to change the mindset of the average UK householder, used to having full control over every aspect of the heat delivered into their homes. With the advent of the Heat Network Assurance Scheme and the 5th generation of heat networks in the UK we are set to see a sea change in delivery, but the big challenge we face remains the change of mindset.

4th generation heat networks reduced the usual circulating network flow temperatures to around 50 – 55°C, making them far more suited to being supplied by central heat pump plant. But the requirement for energy centres, heat pump and in some cases traditional carbon burning boosting technology made it hard to convince the homeowners or developers of the value of these systems. Even with the wealth of experience in the successful roll out of these systems across Scandinavia it has been hard to gain traction in the UK.

The 5th generation heat networks are using ambient loop circuits with integrated district heating and cooling loads, and localised domestic heat pumps delivering temperatures for space and water heating at temperatures which homeowners can more readily identify with. A network flow temperature of between 10 – 30°C is easily achievable with the heat rejected from our nation's cooling plant.

This next generation therefore makes the viability of delivering heat networks utilising waste or recovered heat from alternative sources so much more achievable. Localised heat pumps can use this low-grade heat to deliver space or hot water heating to residences at more traditional levels of supply temperature and give the individual householder direct control over their own system – removing many of the psychological blockers to faster take up of heat pump technology.

And the Heat Network Assurance Scheme and Heat Interface Unit Testing Standard will all achieve a higher level of delivery consistency providing consumer protection and confidence in the technology[7].

## Conclusion

Our sector – representing industrial and commercial refrigeration as well as building services air conditioning and heat pump applications – needs to approach projects with a far more collaborative attitude going forward if tomorrow's buildings are to be truly carbon neutral. Megawatts of usable heat is being rejected from cooling applications every year and much of this could, with a little bit of joined up thinking, be recovered and used to supply heat for heat networks or social housing built near the non-domestic buildings. The building services design consultants need to work more collaboratively with the refrigeration consultants, rather than in the traditional silos, so that we can harness rejected heat and help our built environment become truly carbon neutral.

Integrating the design of building services heating and hot water systems with one eye on the nation's cooling load across so many types of estate, and building into that equation an increased use of renewable energy supplies and a greening of the electrical grid, offers opportunities for rapidly reducing our carbon consumption for our building energy needs.

Every butcher shop, bakery, funeral parlour, corner shop, fast-food outlet has a mixed requirement for space and/or water heating coupled with some specific cooling load, so it is essential that those involved at the design stage of building one of these properties, or even looking at an overhaul of the systems installed at end of asset life, do so with an open mind and incorporate the recovery of the heat rejected from the cooling plant to supplement the heating demands at the very least.

The possibilities have enormous implications for the RACHP sector in terms of increased workload, but this in turn will demand more skilled technicians in our sector. The training requirements for a host of technicians entering the sector or needing upskilling cannot be met by the current way we deliver training. A shortage of trainers and assessors which is not likely to be addressed in the short term means we need to look more holistically at how we deliver upskilling training both for AC and HP engineers through hybrid courses and a different business model for the training centres – with more training delivered as short courses rather than the traditional college delivery approach.

So the opportunities also present several challenges for the sector going forward, but with a more open mind to the commercial opportunities and a more collaborative approach across sectors and even within our own RACHP sector going forward, the future for the sector and our Institute of very bright.

## About the author

Graeme Fox is a Chartered Engineer, a Fellow of the IOR and a Member of CIBSE. He is Director of Technical at the BESA Group, a Director at ACRIB and a Trustee of the IOR, a Past President of AREA, and President of the IOR.

He has over 30 years' experience in the HVAC sector, mainly in building services air conditioning and refrigeration, and for the last 7 years has been working with the trade associations to help contractors with their legal compliance and technical needs, and helping Governments develop policy which is consistent and deliverable.

He has been the driving force in Europe in helping the United Nations Environment Programme to develop minimum competency levels globally to assist the UN Article 5 countries with their transition to alternative refrigerants to meet their Kigali Amendment requirements.

## References

[1] House of Commons Environmental Audit Committee HC 469 2018 *UK Progress on reducing F-gas Emissions Fifth Report of Session 2017–19*

Available from publications.parliament.uk

[2] Department for Business, Energy & Industrial Strategy 2020 UK Greenhouse Gas Emissions, Final Figures https://www.gov.uk/government/statistics/final-uk-greenhouse-gas-emissions-national-statistics-1990-to-2020

[3] IEA Key World Energy Statistics 2015

[4] Committee on Climate Change inquiry November 2020: https://committees.parliament.uk/work/684/technological-innovations-and-climate-change-heat-pumps/#:~:text=The%20Committee%20on%20Climate%20Change,be%20widely%20used%20by%202035.

[5] BSRIA Guide to Refrigerants in Building Services, G Fox https://www.bsria.com/uk/product/rk7vyr/refrigerants_in_building_services_tg_212022_a15d25e1/

[6] Cold Chain Federation: Cold Chain Report 2022

[7] District Heating & Cooling Early Design Building Connections Guide, BESA Publications (currently being prepared by BESA) https://www.thebesa.com/knowledge/shop/products/dhc-district-heating-cooling-early-design-building-connections-guide/

Presented before the IOR on 7 December 2022

## Invitation to join this presentation

The IOR invites you to join this meeting for free in person or remotely – either way we'd like you to be there.

Register your attendance to the presentation on 4:45pm, Thursday 7 December 2022 via:

https://www.eventbrite.co.uk/e/the-rachp-sector-the-critical-part-we-must-play-in-the-net-zero-agenda-tickets-395016915227

Alternatively, you can join remotely by webinar (either by listening live or we can send you a recording of the presentation afterwards) via:

https://register.gotowebinar.com/register/4279657019539422223

The IOR's part in the Net Zero agenda for the built environment - Presidential Address

Gulbenkian Room, One Park Crescent Conference Centre 229 Great Portland St, Marylebone London W1B 1SH

**DISCUSSION REPORT**

**Dermot Cotter,** chairing, thanked the author for a very interesting paper

**Dermot Cotter** – You talk about a system put in 15 years ago and your only concern about the system was that you were using an older heat pump. Why haven't we been doing that for the last 15 years?
**Graeme Fox** – That's a good question. One of the questions was how scalable is this project? I hope that by winning the award recently it will raise the awareness of the possibility for more designers to see that this technology exists. The VRF systems have been used quite widely in the past 15 years. The geothermal side maybe not so much. This is down to cost at the end of the day. The biggest cost in that project was drilling through the granite that the club was built on. The original drilling budget was £30,000 and it actually cost approx. £100,000 because the drilling company were on site for 3 weeks. That was a huge barrier. When they realised, they did ask for alternative suggestions and I suggested air source VRF instead but we couldn't hide them inside the plant room and the aesthetics and architecture were the priority.

**Dermot Cotter** – Should we have a qualification for installing a heat pump into a house because right now anyone can install them and when they go wrong the technology gets a bad reputation? What should be done and who should make the change?
**Graeme Fox** – The industry has been looking at qualifications for that sub sector for the last 18 months. Industry groups have been working with the relevant government departments trying to update the minimum technical competency standards that are in the national standards that were put in place 15 years ago or more. When we sat down to update them it took a long time and after 18 months it's only now that the working groups are going out for consultation. We found a lot of the qualifications are there already but there is a lack of application knowledge that is the problem. The plumbing and refrigeration qualifications haven't changed significantly. It's not so much qualifications that need changing, it's a lack of application knowledge that is the problem and that is why the Heat Pump Association have created new heat pump awareness courses.

**Neil Merritt** - With regards to changes in training, what have you seen as barriers to collaboration with heating and building experts in the industry?
**Graeme Fox** – The biggest barrier has been where people are trying to apply for grants and things like that. In order to apply for grants you need to have MCS accreditation and that is where our sector tends to fall down. It tends be traditional heating engineers that are installer. They don't usually understand what a heat pump is, they don't understand how to properly set up a heat pump. There is some more funding to upskill another 400 or 500 engineers and in conjunction with MCS the BESA qualification that comes from that short course will be MCS approved.

**Dermot Cotter** – Should it be only those people with that kind of qualification are the only ones allowed to install them?
**Graeme Fox** – That's where the upskilling comes in. One of the main features of the course is understanding the different types of system and the applications and the limitations of some systems is very important. Retrofitting is going to be a massive part of the equation but actually retrofitting into old properties, air to water heat pumps are challenging. You need to be delivering a water heating system at a low temperature to realise the energy efficiencies, but in an old building, if you are trying to run that at 40 degrees instead of a traditional 70 degrees then you need to massively upsize the radiators or you could strip out the old ones and install fan assisted ones. The actual pipework itself in old buildings is microbore, unsuitable for low flow temperature heat pump systems so the entire piping system has to come out. Often you are better putting in air to air systems because you don't have all those obstacles, you free up walls that currently have radiators on and also you get the instant heat that you don't get from low temperature. One of the big problems that

households find with an air to water heat pump system is that if they go away they switch the heating off to save energy but you can't then just switch them on when you come back because it will take about a day and a half to actually warm the house again. I read that a BBC journalist has a holiday home and he wrote an article complaining that heat pumps are useless because when he goes for a long weekend and switches it on, by the time it is heating up he is about to go home again. That is not a problem with the heat pump, that is a problem with the design of the system and what should have been installed was an air-to-air system. It would give pretty much instant heat; you don't have to get the heat back into the fabric of the building before you feel the benefit of it. You get the comfort factor pretty much instantaneously. That's the sort of advice that needs to be better communicated.

**Paul Singh** – Which skills, knowledge, behaviours are required to tackle Net Zero?
**Graeme Fox** –The installer needs to understand what they are designing and installing. They need to be able to explain to the householder about the requirement for running at low temperature. The householder is used to having radiators but low temperatures won't give that and, although the system works perfectly well and keeps the temperature warm the mindset of customers is that it's not working because they can't feel heat coming from the system. The actual installers really need to understand the technology so they can properly explain to the clients how to run the systems in the most energy efficient way. That's probably the biggest challenge.

**Steve Gill** – Traditionally heating and cooling have been seen as separate and have been dealt with in silos. How are we going to move away from this thinking?
**Graeme Fox** – Policy makers don't like to be seen to be encouraging the use of air conditioning. Climate change, a warming planet and the equipment that we have in our homes and offices now is putting that necessary cooling load on buildings. That's why I lobbied HMRC for 18 months about heat pumps tax relief because they had removed the reduced VAT rate as air conditioning systems were explicitly written out in 2019. I was able to get them to rewrite their guidance and the Chancellor put in a zero rate of tax for decarbonised heating for 5 years. That was a big hurdle and getting that mindset into government departments that there is a real need for cooling and when that cooling is linked to the most energy efficient heating system that those buildings could have. It's a lot easier and a lot cheaper to install air to air heat pumps in a lot of older properties, particularly housing than it is to strip out and install an air to water system.

**Dermot Cotter** – What would you do in a building that was designed 15 years ago?
**Graeme Fox** – In all honesty I wouldn't do anything differently. I think the data speaks for itself. The equipment has become more efficient, but I wouldn't change the actual design of the system because I think the performance that we have measured speaks for itself.

**Rob Lamb** – On a practical level, you mention heating homes using a heat pump at lower temperatures. What about the hot water for washing etc?
**Graeme Fox** – The electrical grid is greening more and more as time goes on so using electricity to heat water. You could use solar thermal heating and then just use electric resistance heating to boost it. Some of the new generation of heat network equipment is able to deliver water at more like 70 degrees but at an efficient cost.

**Rob Lamb** - As manufacturers and developers what are the key things that we need to be focussing on?
**Graeme Fox -** I think with domestic heating the controls are key. We need really easy controls, they need to be as simple as possible while maintaining performance. That is probably the biggest challenge for manufacturers of heat pump systems.

**Andrew Gigiel** – A few years ago Chris Jessop and myself did some research on heat pumps. You need some sort of energy storage on site or very close to site so that you can disconnect the demand for heat from the demand from the grid. Have you given that some thought.
**Graeme Fox** – I did mention that in the paper about that need for energy storage. It's a really important point. The technology for storage has moved on quite significantly in the last few years. It's become more efficient, but you're not necessarily generating that heat at the time when you will be using it so there has to be some way of storing it.

**Mike Creamer** – I have introduced solar tubes in my home. They produce water at about 75 degrees when the sun is shining but the real issue is at night the sun has gone down and I have no storage. I am wondering if there is some way to combine the two and to use solar tubes to put a lot of heat energy into your water storage and that you could use a water-to-water heat pump in the autumn and winter to lift the heat generated by the solar tubes to higher temperatures.
**Graeme Fox** – There's loads of options available to design engineers and they have to choose the right option for the right application. I haven't talked about solar PV or phase change technology that I think could have a big future in domestic heating and hot water systems.

**Martin Craxton** – What sort of encouragement is there from government for distribution networks and district heating that are going to be the solution for water to water and balancing loads in an urban environment?
**Graeme Fox** – One of the big problems that heat networks have had is poor performance. There is a HIU standard within BESA to address this and most heat interface manufacturers send their units to be tested to these standard now. BESA talked to BEIS about the heat network assurance scheme, and wanted to have a guarantee scheme that gives homeowners the peace of mind that the heat getting delivered by the heat network will be at the right level because that is one of the problems to date is that heat networks have not performed as they were expected so the end user loses faith in the system. That is now progressing. That, I think, is the limit to government involvement. They are not spending money but are helping to progress the heat network assurance scheme.

**Ian Fisher** – We're charging towards Net Zero, yet the global demand for cooling is going to triple by 2030 and in the UK alone the government is demanding a large amount of heat pumps be installed, all needing refrigeration engineers. My question is simple. How many have we got, do we know?
**Graeme Fox** – We have a lack of skilled technicians and we need to upskill. One of the biggest problems we have is that much of our sector don't want to do domestic work so how are we possibly going to meet that requirement? We do have a big problem and that is one of the reasons that I have been looking at helping the BESA academy and helping to develop these upskilling courses because we need to have a huge jump in technicians willing and able to install domestic. We have a huge problem but then every sector has skills shortages, and a lot of the sectors are a lot higher profile than refrigeration and air conditioning.

**Dermot Cotter** – Thank you Graeme.

IOR.org.uk
REFRIGERATION AIR CONDITIONING HEAT PUMPS

The knowledge hub for refrigeration, air conditioning and heat pumps

# From Sustainable Cooling to Heat Recovery - Opportunities and Challenges for Waste Heat in Energy

**Prof Graeme Maidment PhD CEng FInstR**

Winner of the J&E Hall Gold Medal 2022

Copyright © 2023
Institute of Refrigeration

## Why you should attend

- Find out how integration of heating and cooling systems can be used to achieve reduced emissions through smart local energy systems
- Learn from examples of how challenges were addressed, prototypes developed and practical issues dealt with – paving the way for wider roll out of solutions
- Understand the benefits both financial and environmental of waste heat recovery from Datacentres, Supermarkets and Underground trains
- Explore the future potential of using heat from less conventional sources for decarbonisation of the built environment

## Abstract

This paper tracks a research journey over a number of years from sustainable cooling to heat recovery. It includes an investigation of the decarbonisation of heating and cooling by working across commercial boundaries as part of a smart energy system. Many cooling applications exist close to a demand for heat and by sharing energy between applications can deliver near zero carbon heating and cooling in a cost effective way. The paper describes the evolution of research in sustainable cooling and heating. It starts with describing the challenges associated with decarbonisation of the UK energy system and then links this with the need to develop more sustainable cooling concepts. Research carried out on sustainable cooling of London Underground is described including the challenges associated with heat exchange, followed by the evolution of a heat recovery scheme for connection to the Bunhill Heat Network which bivalently cooled the Tube and delivered heating to local housing. The paper then describes the development of smart local energy systems, using a case study called GreenSCIES which integrates heating and cooling with mobility (Electric Vehicles) and power to create a smart local energy system. Wider opportunities for integration of heating and cooling are then described with some specific case studies. The paper concludes by looking ahead to future policy opportunities.

Presented before the IOR on 2 March 2023

## 1. Introduction

In 2019, the United Kingdom became the first major economy to commit to a legally binding target of cutting greenhouse gas emissions (GHG) to net-zero by 2050 (BEIS, 2021a). This target was motivated by the increasing uptake of renewable energy sources for electricity production in the UK, with renewables reaching an all-time record of 37% of the power generated in 2019 (BEIS, 2020); and the UK Government stated that it would fully decarbonise the electricity generation system by 2035 (UK Government 2021).

The decarbonisation of the electricity grid, makes the electrification of heat, e.g. through the use of heat pumps, a great opportunity to help reduce the carbon footprint of UK buildings, as recognised by the UK Government's Heat and Buildings Strategy (BEIS, 2021b). This assumes that the repurposing of the gas networks to run hydrogen boilers doesn't materialise as stated in the Chris Skidmore review of UK Net Zero (Skidmore, 2023).

However, heat pumps will put specific demands on the electricity system. Achieving net zero through a decarbonised electricity presents a number of challenges but with it, opportunities. The challenges are:

- The impact of increased electrification on the grid and the potential cost associated with strengthening it. Increased use of electric vehicles and heat pumps potentially doubling demand for peak electricity.
- The balance of supply and demand.

Challenges can be overcome by both minimising demand through energy efficiency measures but also by using electricity flexibly. Heat is responsible for nearly half the energy consumption and a third of carbon emissions in the country (BEIS, 2021) and heating and cooling have been previously regarded as being hard to abate. However, as significant energy users and carbon emitters and as often termed flexible assets, heating and cooling systems present a great opportunity for reducing emissions through energy efficiency and as well as through flexibility as part of a smart local energy system.

Cooling and heating systems are now part of a complex energy system but it wasn't always this way. This paper charts the evolution of research from sustainable cooling, to cooling and heating to being part of a complex smart local energy. It starts with describing sustainable cooling of London Underground including the challenges associated with heat exchange followed by the evolution to the Bunhill scheme which bivalently cooled the Tube and delivered heating to local housing. The paper then describes the development of smart local energy systems integrating heating and cooling with mobility (Electric Vehicles) and power to create a smart local energy system through a case study called GreenSCIES. Wider opportunities for integration of heating and cooling are then described with some specific case studies. The paper concludes by looking ahead to future policy opportunities.

## 2. Sustainable cooling on the London Underground

Cooling of the London Underground is one of the long standing challenges for London. Our work in this area began in 2001 and we soon discovered that it is not straightforward to achieve because of a number of factors that impact on cost to implement. Many of these are a consequence of the age of construction of the Tube (much of which was constructed in the mid c19th). These factors include:

- Space on trains to fit heat exchangers and air conditioning.
- Space constraints in tunnels to fit heat exchangers and pipework.

- The lack of ventilation to reject heat from the tunnels to outside (ventilation rates are about 1/10th of those for the Hong Kong Metro).
- Safety – This is paramount everywhere partly because of the Kings Cross disaster in 1996 and quite rightly it makes any complex intervention on the Tube much more complicated.

## 2.1 Heat on London Underground

The low ventilation rates over many years combined with increases in the frequency of trains due to the increases in acceleration and deceleration across the network has resulted in significant temperature rises over time on the Tube as shown in Figure 1. Temperatures of more than 37°C have been reported on some trains in summer (The London Paper, 2009 and Evening Standard 2003). In the record breaking 2022 heat wave temperatures of only 36°C were recorded (Evening Standard, 2023), although this may be due to the number of passenger journeys during 2021-2022 still being lower compared to pre-COVID-19 pandemic numbers.

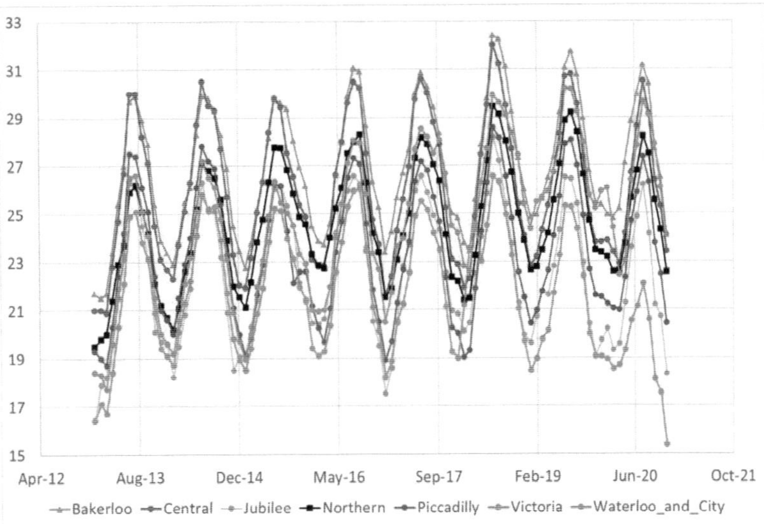


**Figure 1 - Temperatures on the Tube. (Adapted from TFL, 2021)**

Extreme heat and cooling of the London Underground was the starting point for much of our research work today. Our initial work centred on identifying the heat loads in both tunnels and trains in order to be able to mitigate these (Ampofo et al, 2003). We found that most of heat in the train was generated by the individuals using it and 85% of heat generated in the tunnels and platforms was generated by resistance braking as shown in Figure 2 below (Gilbey & Thompson, 2009). Both train and tunnel/ platform heat loads are/ were unavoidable without effectively changing the train stock or running with lower ridership. Unfortunately both options either impacted on the business or were overtly expensive and therefore not realistic solutions.

Figure 2 – Thermal image showing heat generated from braking underneath trains.

## 2.2 Groundwater cooling concept

Based upon the above we began to consider cooling possibilities. However, it was soon apparent that cooling of London Underground is not straightforward, and to overcome challenges we proposed an alternative approach to delivery of cooling. The concept is shown in Figure 3 and involved the use of groundwater as a cooling source. As shown, groundwater was abstracted from the aquifer and circulated through an air handling unit before returning to the point of abstraction. The concept worked well because groundwater is at an appropriate temperature and scale for cooling the Tube, and it is at a convenient location for use. Indeed, in several cases we established that ground and seepage water was already being pumped out to protect the London Underground infrastructure, so as well as being able to deliver cooling in a more sustainable way, the concept also protected the infrastructure – a further benefit.

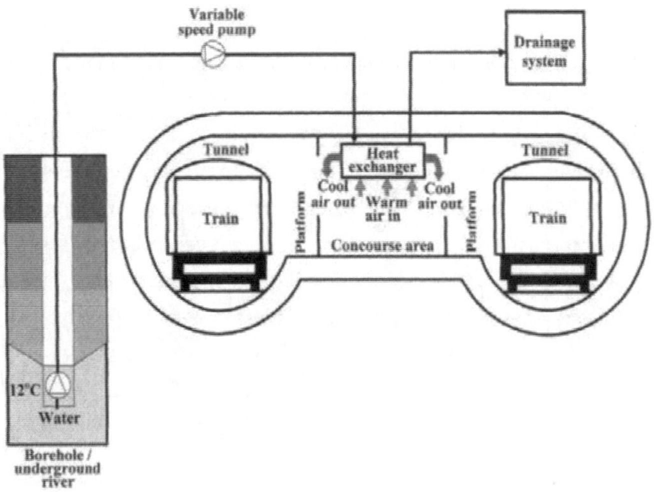


Figure 3 – The proposed groundwater cooling concept

The idea was published at a CIBSE conference in July 2001, the day before 40 members of the public were stuck on the Tube during periods of extreme hot weather. The project was reported on the 4th of July 2001 in the Evening Standard as shown in Figure 4.

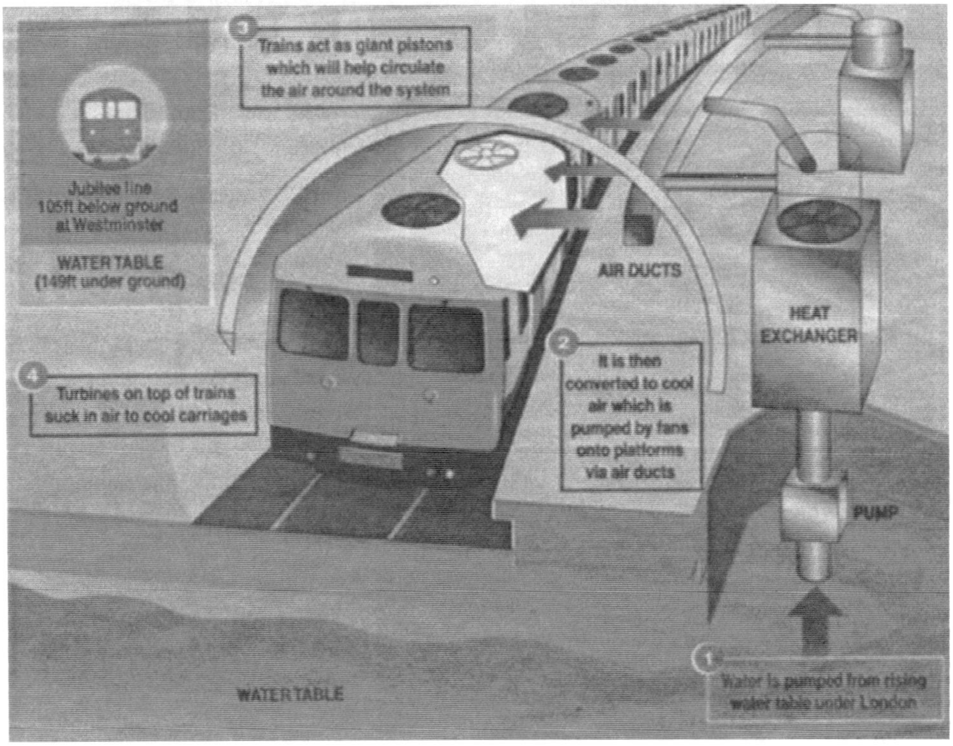


**Figure 4 – System as published by the Evening Standard (2001)**

Over the next few years the concept was developed which included gathering data on water pumping rates, water quality and temperature. The data indicated that the scheme was potentially feasible and proceeded to a feasibility trial.

### 2.2.1 Trial at Victoria

The trial site chosen was Victoria Tube station because groundwater was already pumped out of the station. The system was designed to supply 150 kW of cooling. This was achieved by using three 50 kW fan coil units of air inlet and outlet dry bulb temperatures of 32°C and 22°C, respectively, and 50% relative humidity and water inlet and outlet temperatures of 16°C and 22°C, respectively. The system was sited at the concourse area between Northbound and Southbound Victoria line platforms at Victoria Tube station and Figure 5 shows one of the installed fan coil units. The prototype installation was completed in July 2007 and fully commissioned at the beginning of August 2007.

**Figure 5 – The Victoria Groundwater trial (2007)**

The prototype system at Victoria Tube station was then monitored for twelve months after commissioning. The aim of the monitoring and testing was to investigate its performance and capabilities with the aim of an optimal design allowing the technology to be rolled out to other parts of the London Underground network.

The specific objectives of the monitoring and testing were to:

- Measure operating parameters from the cooling system,
- Measure environmental parameters from within Victoria Tube station,
- Investigate the impact on thermal comfort conditions,
- Estimate cooling system output, efficiency and coefficient of system performance (COSP).

Figures 6 and 7 shows some results from the trial. Figure 6 shows the flow and return water temperatures for the fan coil units (FCU) for 01/08/2007 – 31/08/2007. In the figure, "FCU1 return", "FCU2 return" and "FCU3 return" are the average water temperature data recorded for the three fan coil units, respectively; FCU1 is shown in Figure 5. Also shown in Figure 6 is the air temperature data at two locations within the concourse area (Lower escalator landing and Intermediate concourse) and at the exterior of Victoria station. Figure 7 shows the cooling system COSP for August 2007, based upon pump efficiencies of 50% and 70%. It can be seen that if a reasonably efficient pump is used, that an average COSP of 16 is achieved.

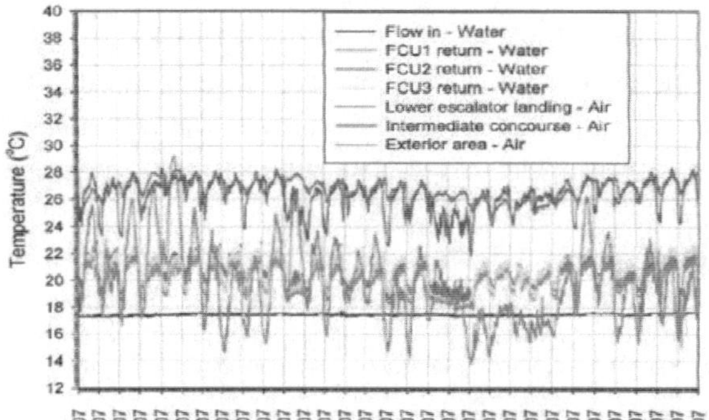


**Figure 6 – The Victoria Groundwater trial temperature data (2007)**

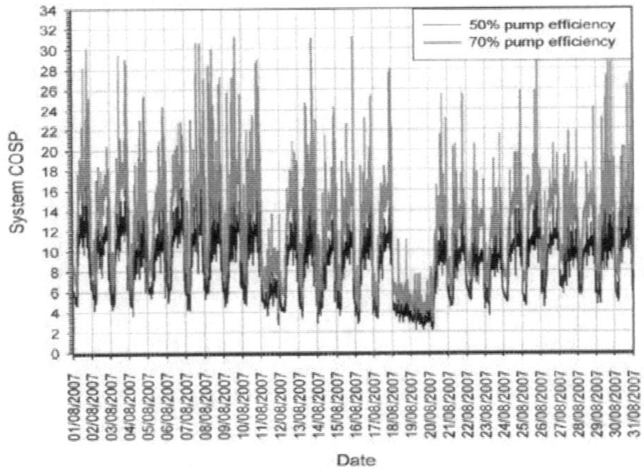


**Figure 7 – The Victoria Groundwater trial system COSP data (2007)**

After around 1 month, the performance of the fan coil units degraded. This was caused by the blockage of filters at the FCU inlet due to particulate build up. The filters were cleaned and performance restored, however, the performance degraded again after a similar period. A picture of the degraded filter is shown in Figure 8.

**Figure 8 – Filters from the Victoria Groundwater trial (2007)**

### 2.2.2 Trials at Green Park and Bond Street

Due to the recession in the late 2000s, the work was shelved, however, with the Olympics of 2012 it was decided by TfL to install platform based cooling schemes at Green Park and Bond Street stations. Green Park used the groundwater concept as its cooling source, whereas the Bond Street Station used an external chiller. Both schemes used Platform Air Handling Units (PAHUs) as shown in Figure 9 to cool air on the platforms, which included finned tube heat exchangers and fans, similar to those tested at Victoria Station. Filters were not used to protect the coils from fouling, but instead an anti-fouling coating was added.

Both schemes were tested after installation and were deemed partially successful in delivering cooling to the platforms. The reason that they were only considered partially successful was that the coil performance degraded quickly due to fouling . We also understand that the PAHU's were difficult to clean on site and to remove and clean offsite required the use of an engineering train which needed to be booked 6 months in advance and cost £100,000.

**Figure 9 – PAHUs installed at Green Park Station in 2012**

The challenge of keeping heat exchangers clean as well as the cost of installing the cooling system were two contributory factors which lead to future research for the LSBU team, as detailed below.

## 2.3 Overcoming Particulates: barriers to roll out

This section describes the work carried out to overcome the barriers associated with reducing the cost of implementing the use of heat exchangers in dirty environments.

### 2.3.1 Heat exchange in dirty environments

Early on in the Victoria trial we realised that particulates have a massive impact on the performance of the heat exchanger. Unfortunately, lessons learned were not fully adopted into further trials and in 2015 TfL invited LSBU to get involved again in the project and specifically to support a PAHU trial to investigate alternative heat exchanger configurations.

The purpose of the project was to identify coil configurations capable of delivering cooling on the Tube. Specific objectives including:

- Identifying design options which reduce the maintenance requirements of the units and avoid the need for regular track access, whilst maintaining the high cooling performance, thus providing significant cost savings to London Underground.
- Studying differences between dust accumulation patterns on various heat exchanger coil configurations and the resultant impact of the air side fouling on the coil heat exchanger performance.

The project investigated six different PAHU design concepts which were installed in a disused station. This was on an operational line and therefore ensured testing of the AHUs performance in an operational LU environment, akin to the environment in which they would be required to operate in future. Measurements were taken to monitor the performance of the units as well as the rate of build-up of the dust.

The location of the trial was critical to ensure that the environment was representative of the AHUs location in a station. After careful consideration based on ease of access for installation, tunnel temperature, ability to collect data during the day, and the airborne environment, a disused station (on an operational line) was chosen. The trial was set up in the station at platform level, in an area where the platform used to be. This is shown in Figure 10.

**Figure 10 – PAHUs being tested in 2015**

In total six PAHUs were developed and tested, with different combinations of filters. These were then compared to a control unit based on the design of the existing units installed and operational at Green Park and Bond Street stations. Table 1 shows details of each unit in terms of the coil and fan configuration, whereas Table 2 shows the range of filters tested.

| | AHU 1 - Control | AHU 2 | AHU 3 | AHU 4 | AHU 5 | AHU 6 |
|---|---|---|---|---|---|---|
| Coil Type | Rippled Fins - 8 fins/inch, 12 rows | Flat Fins - 4 fins/inch, 12 rows | Flat Fins – 4 fins/inch, 12 rows | Flat Fins - 4 fins/inch, 8 rows | Flat Fins 4 fins/inch, 12 rows | Flat Fins 4 fins/inch, 12 rows |
| Fan | Axial fan, fixed speed | Axial fan, external VSD | EC plug fan, integral speed control | Axial fan, fixed speed | EC plug fan, integral speed control | EC plug fan, integral speed control |

**Table 1 Details of the AHUs Tested**

| Type of Filter | Filter Class based on EN779:2012 | Filter type |
|---|---|---|
| Metal Mesh Panel | G2 | Coarse |
| Metal Mesh Panel | G3 | Coarse |
| Delbag Panel | G4 | Coarse |
| Delbag Bag Filter | G4 | Coarse |
| AAF Panel Filter | M6 | Medium |
| AAF Bag Filter | M6 | Medium |

**Table 2 Details of the filters tested**

The experiment derived cooling capacity (thermal performance) by calculating the heat transfer on the water side. Filter and coil degradation (hydraulic performance) were measured using the pressure drop across the coil and filter respectively. Results showing the increase in pressure drop and degradation of each PAHU is shown in Figure 11.

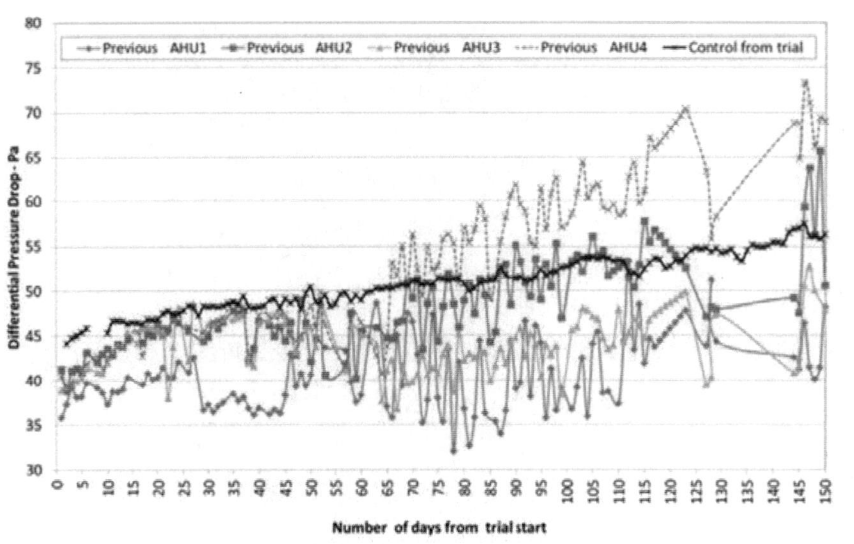


**Figure 11 – Pressure drop recorded across PAHUs from the start of the test period**

| PAHU number | Filter and coil arrangement | Coil Arrangement | Reduction in coilcapacity in 3 months | Filter Type | Filter Class | Duration between filter changes (weeks) | Comments |
|---|---|---|---|---|---|---|---|
| 1 | no filter + | Rippled Fins, 8FPI, 12 rows | No noticeable reduction | None | None | | The control unit |
| 2 | + | Flat fins, 4FPI, 12 rows | Approximatley 20% reduction | Metal Mesh Panel | G2 | 17+ | Removed metal particles but not fibrous material |
| 3 | + | Flat fins, 4FPI, 12 rows | No noticeable reduction | Metal Mesh Panel | G3 | 3 | |
| | + | | | Delbag Bag Filter | G4 | 5.3 | |
| | + | | | AAF Panel Filter | M6 | 0.75 | |
| 4 | no filter + | Flat fins, 4FPI, 8 rows | No noticeable reduction | None | None | | Unprotected coil with wide fin spacing |
| 5 | + | Flat fins, 4FPI, 12 rows | No noticeable reduction | Delbag Panel Filter | G4 | 4.5 | |
| 6 | + | Flat fins, 4FPI, 12 rows | No noticeable reduction | AAF Bag Filter | M6 | 5.4 | |

**Table 3 – Summary of results**

Table 3 shows a summary of the initial results for the six AHUs configurations over approximately a three month period. As can be seen PAHUs 2, 5 and 6 have been tested with one specific filter configuration, PAHU 3 has been tested with three different filter configurations and PAHU 1 and 4 did not have any filter fitted. Initial results showed that fouling had only a small impact on heat transfer with the exception of AHU 2. On inspection of the PAHU 2, it was observed that this was caused by fouling acting as an insulation layer.

Table 3 also shows the performance of the filters over three months. This shows that with the exception PAHU 2 filter, no filter could achieve a ten week duration between filter changes, with a maximum of 5.4 weeks being achieved. This is too short and would require many filter changes over the season which would be operationally and financially prohibitive for LU. Therefore the use of these filters is precluded. By examining the pressure drop across the unfiltered coils we found that the pressure drop across PAHU 1 increased, whereas the wider fin spacing of PAHU 4 did not.

The fouling mechanism of PAHUs 1 and 4 was studied qualitatively by visually recording the build of material over time. This is shown in Figures 12 and 13. Figure 13 shows the build up with the control unit with narrow fins (PAHU 1). As can be seen fouling begins to have an impact when bridging across the fins occurs. This can be caused by either fibres or bridging taking place at the shoulders of coil tubes, as indicated. However, with PAHU 4 (as shown in figure 13) with wider fin spacing, bridging at the collar does not occur and build-up of particles is prevented as the coil appears to "air wash" or self-clean once a specific thickness of particles is achieved. Bridging of fins can be caused however by fibres, such as the Dandelion seed shown in Figure 13.

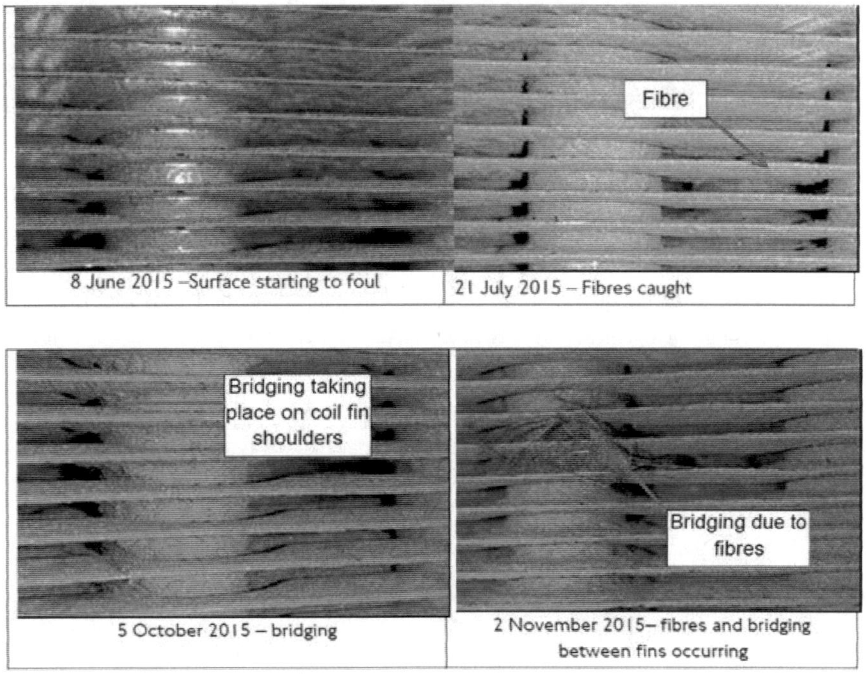


8 June 2015 –Surface starting to foul | 21 July 2015 – Fibres caught

5 October 2015 – bridging | 2 November 2015– fibres and bridging between fins occurring

**Figure 12 - The Fouling Process PAHU1**

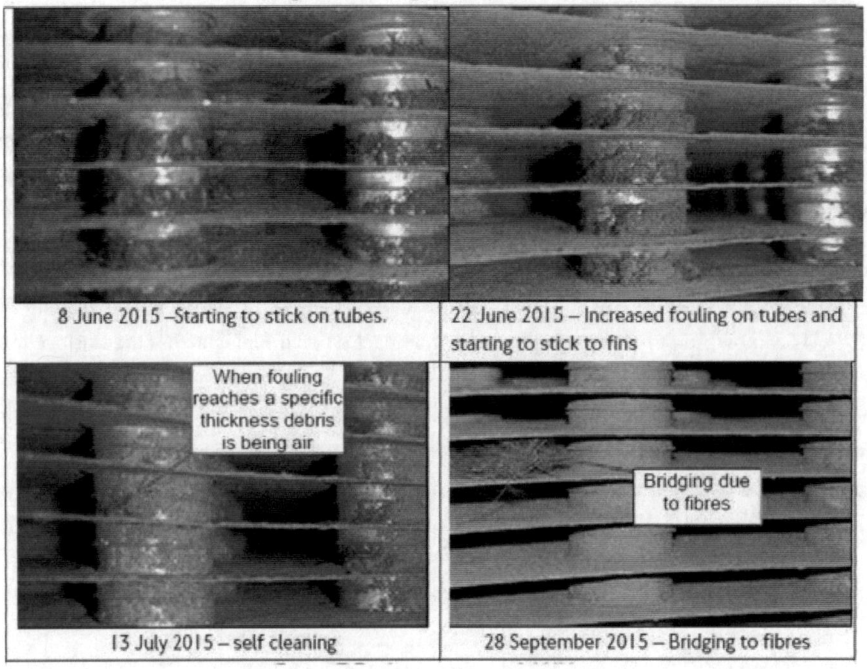


**Figure 13 - Fouling Process of PAHU4**

### 2.3.2 Cost

The second biggest challenge to implementing widespread cooling was the capital cost. A particular challenge has been the deep Tube lines i.e. Bakerloo, Central, Jubilee, Northern, Piccadilly, Victoria and Waterloo & City lines. Doing anything new on the Tube is challenging cost wise due to the risk averse nature of working on the Tube, extensive safety considerations and the need to work engineering hours (1am-5am). Engineering is even further complicated on the deep-level Tube lines, as there are space constraints with smaller trains running through two circular tunnels with a diameter of about 11 feet 8 inches (3.56 m). It has been widely reported that it is challenging to lower temperatures on the deep Tube lines, as traditional cooling systems have proved prohibitively expensive and difficult to install within the 120-year-old tunnels and stations (TfL, 2022). A capital expenditure figure that has been often socialised is £1B / degree of cooling.

To make the economics more favourable Thompson et al (2006) proposed the recovery of heat from the underground for use externally for heating and hot water. By utilizing the heat in this way provides an additional income stream at marginal additional cost and thus improves the economics. Bivalent heating and cooling is the main thrust of the rest of the paper.

## 3. Combining cooling and heat recovery from the Tube.

Integrating heat recovery and cooling from the Tube was originally investigated as part of a project called MICAH (Metropolitan Integrated Cooling and Heating) funded by Innovate UK. It was developed in parallel with the Bunhill Project as part of an EU project "Celsius" (Celsius, 2020). Both of these projects are described below:

### 3.1 The MICAH concept

The main premise for the Metropolitan Integrated Cooling and Heating (MICAH) concept is that the significant quantities of waste heat generated by London Underground are currently dissipated using ventilation shafts, which are located at intervals of e.g. 1 to 2 kilometres, along the tunnels. The ventilation shafts, which incorporate fans, may be operated in either forced ventilation (or supply) mode, whereby ambient air is supplied to the tunnels, or in draught relief (or exhaust) mode, whereby the heated air is exhausted from the tunnels.

The MICAH project involved investigating the potential benefits and barriers involved in implementing a system for combining cooling of a London Underground (LU) site with waste heat recovery (Davies et al. (2019). This system involved the use of a fan coil heat exchanger located close to the head of the shaft, incorporating a reversible fan, enabling its use in both supply and exhaust modes. In each case, heat would be extracted from the air passing over a cold water, heat exchanger coil, whereby the heat would be transferred to the water, raising its temperature. The heated water would then be upgraded using a heat pump (HP) before delivery to a heat network at a typical temperature (for conventional domestic water radiator heating systems) of 70°C.

A schematic of the proposed system is shown in Figure 14.

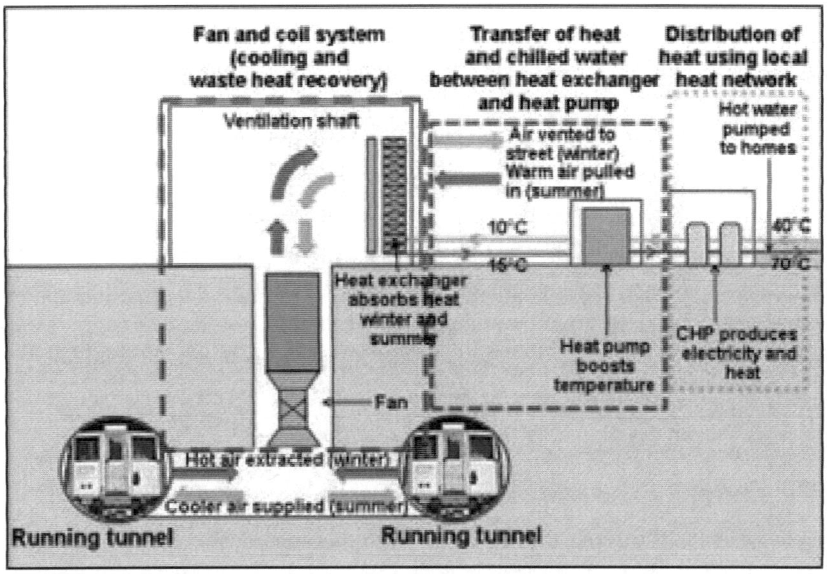


**Figure 14 - Schematic of MICAH system**

By using a reversible fan, the ventilation shaft could operate in supply mode to provide cooling during the summer months, and in extract mode during the winter months, when cooling is not needed.

To evaluate the performance of the system, a Microsoft Excel spreadsheet model was developed which simulated the heat transfer processes comprising the MICAH system, and linked the key input parameters to the performance output parameters, namely: (1) the recovery of heat from the air passing over the fan coil heat recovery heat exchanger (HRHX); (2) a primary water loop which pumps the heated water exiting the HRHX and transports it to the heat pump evaporator; (3) upgrading the temperature of the recovered heat using a heat pump; (4) a secondary water loop for transferring the upgraded heat to the DHN. A schematic of the model showing typical operating temperatures is shown in Figure 15.

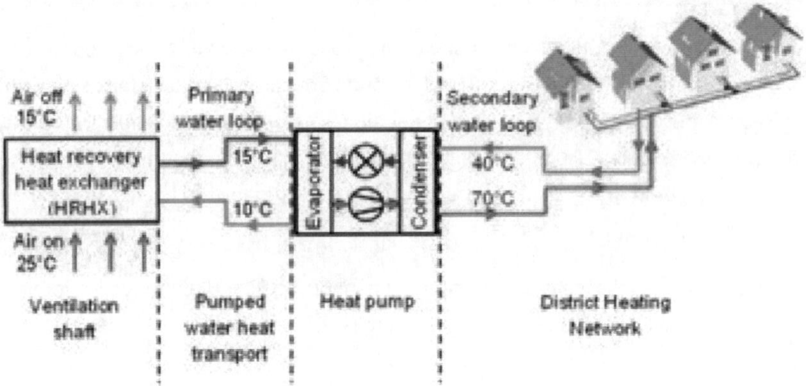


**Figure 15 - Schematic of MICAH model**

Two different layouts for the system were considered, namely one where the heat pump was located in the ventilation shaft head house close to the fan coil heat exchanger (layout 1), and a second where the heat pump was located in the district heating network energy centre at a distance of 840 m from the ventilation shaft (layout 2).

Based on the electrical energy input for both the heat pump and pumping for the water loops, the corresponding operating costs and CO2e emissions for MICAH were calculated by the model and compared with those resulting for delivery of a similar amount of heat and cooling using conventional methods, namely heating from a CHP system and cooling using a vapour compression refrigeration (VCR) system. The results are shown in detail in Davies et al. (2019) but extensively showed significant savings in terms of energy use, carbon emissions and costs, compared with CHP heating and VCR cooling. With 200 ventilation shafts in London, it is predicted that it would be possible to save annually more than 360,000 tonnes of carbon and generate revenues of £40,000,000, compared to conventional cooling and heating systems. If this technology can be applied in 50% of 150 metro systems worldwide (UITP, 2015), more than 27 million tonnes of carbon savings will be achieved. This is expected to increase further as the electricity grid continues to be decarbonised, as the proportion of renewable energy sources increases, in the future.

### 3.2 The Bunhill Study

The opportunity for combined cooling and heat recovery, further consolidated by the MICAH concept, supported the development of the Bunhill 2 project. This was an expansion project for the Bunhill Heat Network in Islington, which introduced a first-of-its-kind scheme that recovers waste energy from a ventilation shaft of the LU network as a novel heat source. The expansion added six new buildings to the district heating network (DHN), which now delivers heat to 1,350 dwellings, two leisure centres and a local primary school.

The project also involved constructing a new energy centre at the City Road ventilation shaft, located on the Northern Line between Angel and Old Street stations in Central London. This is a disused station from the 1920s from which waste heat is now being recovered. The new energy centre houses the waste heat recovery (WHR) system, a 50 m$^3$ thermal store and two 237kWe/372kWth combined heat and power (CHP) units. The main components of the WHR system are heat recovery coils with a nominal capacity of 780 kW, a 1 MW two-stage ammonia heat pump, and a reversible fan, which allows the system to operate in the Extract and Supply modes proposed in the MICAH concept. In order to accommodate the heat pump and allow for a more efficient operation, network temperatures were reduced from 95/75°C to 75/55°C. Photos of the heat recovery coils and the heat pump are provided in Figure 16, and a schematic of the Bunhill 2 energy centre is provided in Figure 17.

(a) (b)

**Figure 16 – Photos of the heat recovery coils (a) and heat pump (b) at the Bunhill 2 energy centre**

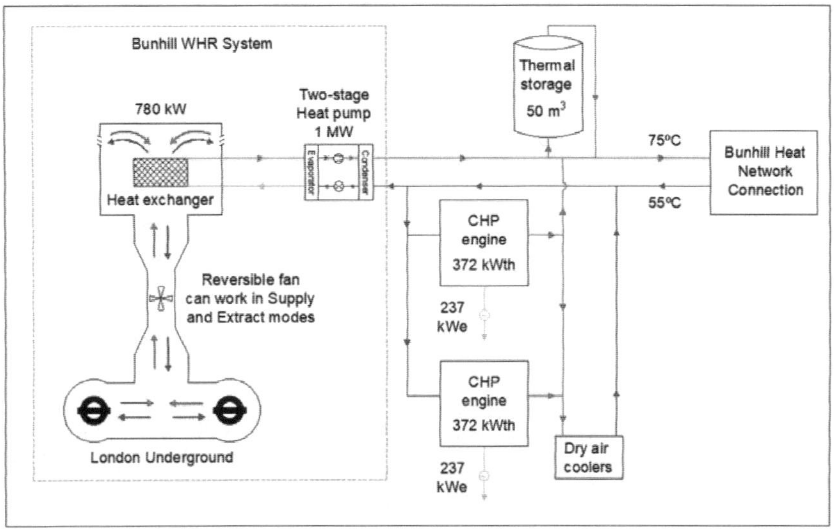


**Figure 17 – The energy centre for Bunhill 2, highlighting the WHR system and its main components.**

Building upon previous collaborations, LSBU partnered with London Underground to evaluate the performance of the WHR system against conventional heating technologies, and investigate potential impacts on the LU tunnelled environment when operating in Supply mode. This was achieved by developing a WHR model able to iteratively solve thermodynamic balance equations across the system, being used to determine its energy consumption, as well as heating and cooling outputs (Lagoeiro, 2022). This model was then coupled with a bespoke modelling tool used by London Underground in order to simulate how future network temperatures would be affected by the WHR system, as shown in Figure 18(a).

These models were applied to investigate how different operating conditions, involving both Extract and Supply modes, would affect the potential benefits that could be obtained by recovering waste heat from the London Underground. Firstly, the WHR system was evaluated based on its carbon abatement costs, in £/tCO2e, for different operating modes (Lagoeiro. 2021), and these were compared with ASHPs using gas boilers as the reference case. The results of this investigation are summarised by Lagoeiro. As a summary however, operating in bivalent, with 6 months in each mode (6E/6S), would reduce decarbonisation costs by 18% compared to ASHPs. This is due to the energy efficiency gains from a bivalent operation that can supply both heating and cooling, which would increase the COP for the WHR system to 5.4, as opposed to a COP of 3.4 that would be achieved when operating only in heating mode (i.e. Extract Mode).

| Scenario | Fan Operation | Description |
|---|---|---|
| 1 | 12E/0S | Fan operating in extract mode for the entire year (12 months). |
| 2 | 9E/3S | Fan operating in supply mode during meteorological summer (Jun/Jul/Aug), and in extract for the rest of the year. |
| 3 | 6E/6S | Fan operating in supply mode for half the year, from May to October, and in extract for the remaining 6 months. |
| 4 | 3E/9S | Fan operating in extract mode only during meteorological winter (Dec/Jan/Feb), and in supply for the rest of the year. |
| 5 | 0E/12S | Fan operating in supply mode for the entire year (12 months). |

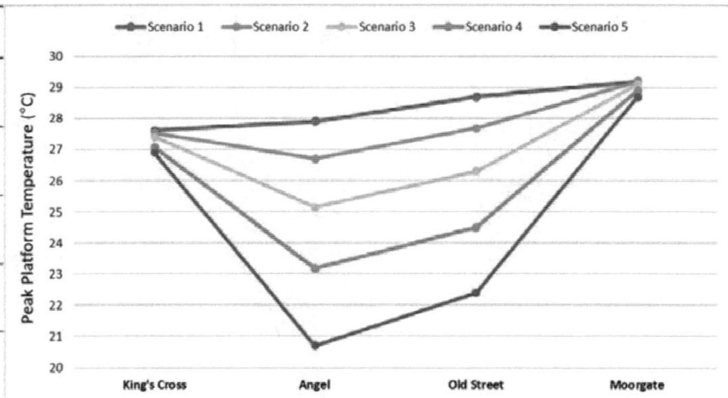


**Figure 18 – Peak platform temperatures for 2030 based upon a combination of extract and supply mode simulations**

Figure 18 indicates that the WHR system could also have a significant benefit in terms of alleviating future peak platform temperatures, with reductions of up to 7.2 K being projected for the adjacent Angel station in 2030. In a 6E/6S operation, ΔTs of up to 3 K could be achieved. The results of the investigation have identified a great opportunity to combine heating and cooling when exploiting waste heat, thus significantly increasing energy efficiency and potential benefits.

## 4. GreenSCIES and Smart Local Energy Systems

Smart local energy systems provide an opportunity to integrate the provision of heating/ cooling, mobility (electric vehicles) and power into a local energy system. By integrating heat power mobility into a common system it's possible to increase further the value streams, share costs and therefore increase financial viability. Smart local energy systems (SLES) have the opportunity to revolutionise how energy is used in urban areas for the provision of Mobility (EV), Power and Heat. The development of smart local energy systems was stimulated by Innovate UK's PFER (Profiting from the Energy Revolution) programme (Innovate UK, 2019).

This originally funded some feasibility studies and demonstrations as well as concept design projects. One of which GreenSCIES is described below.

## 4.1 The GreenSCIES concept

GreenSCIES is a smart local energy concept that has been developed with the aims of reducing carbon emissions and tackling fuel poverty originally across the London Borough of Islington (LBI). The initial project aimed to help Islington Council (IC) achieve its ambition of being a net zero carbon borough by 2030. The core concept is an ultra-low temperature 5th generation heat network enabling recovery of low-grade waste heat such as from data centres and the London Underground. It also includes distributed energy centres with low carbon heat pumps supplying heating/cooling from the ambient loop to local buildings. Each of the energy centres will provide hubs for photovoltaic (PV) electricity generation, electric vehicles and vehicle-to-grid charging/storage alongside large scale batteries. The hubs can then be used for Demand Side Response to flex with the electricity grid requirements/tariffs using a sophisticated artificial intelligence control system. A schematic showing the GreenSCIES concept in Islington is detailed in Figure 19 below.

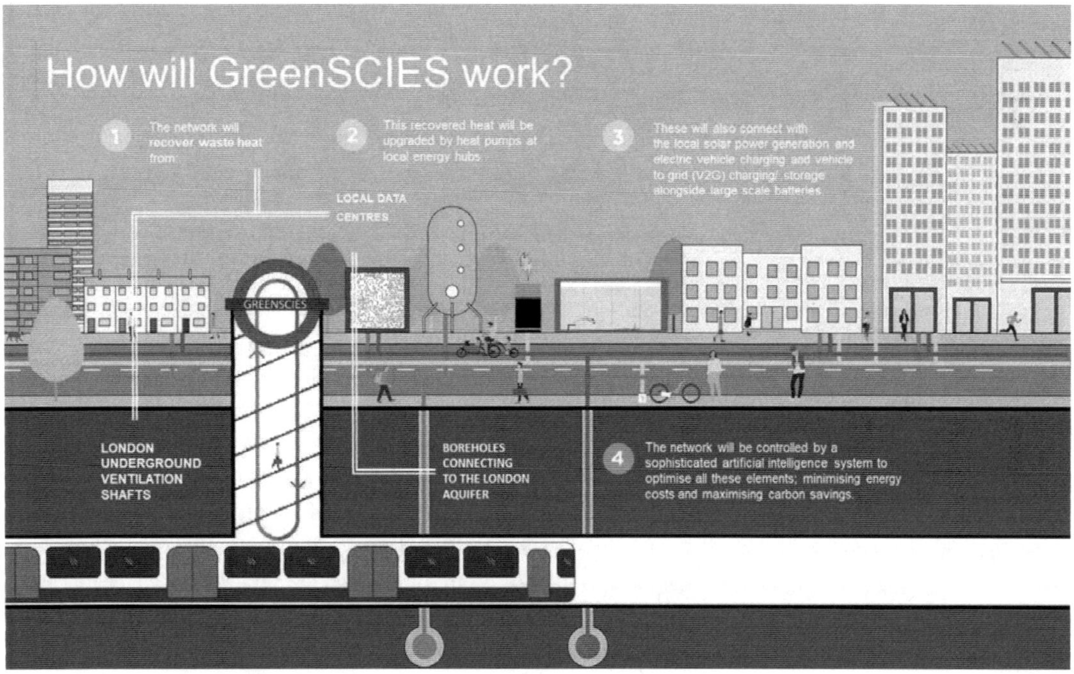


**Figure 19 - The GreenSCIES Concept.**

## 4.2 GreenSCIES in Islington

In Islington, the GreenSCIES concept has been developed into a detailed design called the "New River" scheme which is shown in Figure 20 below. The scheme is built around a data centre from which we recover the heat and in doing so deliver around 4 MW of cooling. This heat is then transferred via a 5th generation ambient loop network to 7 energy centres in close proximity which include heat pumps to use and amplify the temperature of the recovered heat for heating and hot water. Each energy centre includes thermal storage to give flexibility potential. The scheme connects to 3 social housing blocks, one school, one university (with multiple buildings), a library and a theatre. In the university, the scheme also connects into condensers of chillers to recover heat and improve the efficiency of the chillers.

It also includes around 50 electric vehicle (EV) charging points and close to 1 MW of PV. This is integrated behind the meter and therefore avoids the green tariffs applied to grid electricity.

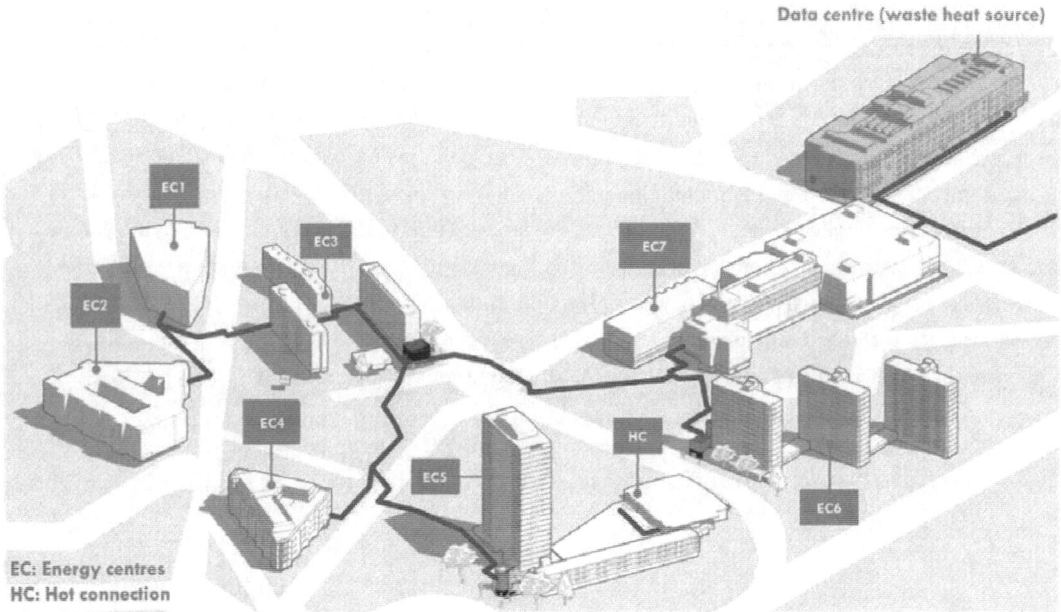


**Figure 20 - Schematic showing the arrangement of the New River Scheme.**

The scheme was originally investigated as a feasibility study and subsequently was developed into a detailed design. This scheme developed result was shown to achieve an 80% reduction in emissions compared to the traditional mobility, power and heating systems it would replace and has the potential to achieve an IRR (Internal Rate of Return) of up to 10% (Revesz et al., 2019).

The positive IRR was due to the value streams that the scheme combined. These included the delivery of cooling as a bi-product of heating and the subsequent income streams associated with both. EV charging was installed at the same time as the ambient loop sharing capital cost and thus minimizing initial expenditure. There were also mobility income streams derived from EV charging which benefited the scheme's viability and increased the IRR by 1%. By utilizing PV power behind the meter means that the end user can benefit from lower cost electricity. There are also opportunities to benefit from financial flexibility incentives since heat pumps and EVs are flexible assets and therefore reduce OPEX costs. Flexible use of the assets would allow the scheme to pick the best electricity price periods as well as accessing the capacity market such as firm frequency response (FFR) balancing mechanisms. To control the scheme we developed a control algorithm enabling dispatch of heating/ cooling/ EV assets in response to demand and price signals from the grid.

As well as the savings associated with cost and carbon, there are other benefits including social value i.e. the creating of local jobs, reduction in particulates etc. Islington Council are currently developing into a programme to potentially construct. If constructed this will be the first large smart energy system in the UK that integrates energy technologies across heat, power and transport.

### 4.3 Other GreenSCIES Case studies

The GreenSCIES concept is potentially replicable in any urban area depending on the local characteristics. Critical to the success is having a potential accessible secondary waste heat source available to recover heat from whilst delivering cooling. The GreenSCIES consortium has investigated GreenSCIES concept feasibility studies in 7 other locations across the UK. These have included utilising heat from data centres, supermarkets, glass works, a foundry, an incineration plant and mine water. One of these applications is described in more detail below and involves the integration of the GreenSCIES concept around a supermarket in an urban setting. By integrating supermarkets into local energy infrastructure they can share resources and provide local communities with low carbon, low cost decarbonised sources of energy.

### 4.4 Supermarkets as energy hubs

Supermarkets are large users of energy and are also large emitters of carbon, much of which is generated through the operation of heating and cooling. Supermarkets and UK households rely extensively on gas for heating. Removing the need for gas heating has significant decarbonisation potential. Supermarkets also have a big cooling requirement and in doing so reject a significant amount of heat. If we view a supermarket as a waste heat source and nearby residential developments as heat demands, then there is clearly waste heat recovery opportunity. The integration of a GreenSCIES SLES approach based around a 5th generation heat network with a supermarket has significant potential to reduce carbon emissions. Decentralised heat pumps at the end-user sites within 5th generation networks (ambient temperature) can be used to boost the temperature of the water in the low temperature distribution loop to provide space heating and hot water for end users. A low temperature heat recovery loop can supply water source heat pumps that can raise temperatures to the building requirements. The use of local thermal storage can help to smooth heat peaks and balance supply with demand. In addition to having a secondary heat source, supermarkets have attributes that make them particularly interesting as an energy hub. Supermarkets could utilise their thermal assets to provide flexibility to the grid. In addition, they have significant parking facilities that could be used for electric vehicles (EVs) and have large roof areas that can be exploited for renewable energy technologies. Supermarkets are often central to the local community, are close to a range of different end users and potential prosumers of heat, power and mobility.

To explore the potential the GreenSCIES team carried out a feasibility study on a real existing store in an urban setting. The store is shown in Figure 21 and a schematic showing how the refrigeration system connects to a heat pump via an ambient loop is detailed in Figure 22. Figure 23 shows a close up of the piping routes.

**Figure 21 - Photograph of the supermarket showing heat recovery, heat pump and new residential development.**

Figure 21 shows heat being recovered from 7 roof mounted refrigeration pack condensers. The low grade heat is then supplied to a new Energy Centre (in red) where a 750 $kW_{th}$ water source heat pump raises the temperature to 65°C to supply the new residential development to the rear of the picture. The development is now complete and comprises some building renovation and new build providing a total of 284 properties. It is understood that space heating and hot water are provided via a communal heating system supplied by central gas boilers. Heat from the new large heat pump would be supplied via Plate Heat Exchangers to interface into the communal heating and thereby displace gas boiler usage. A new thermal store would be included in the energy centre to smooth the operation of the heat pump and balance supply with demand. Displacing gas use in the supermarket itself was also included in the modelling initially but this was rejected because of the difficulty and cost in conversion of dispersed air handling units which currently use direct gas fired heaters rather than LTHW heater batteries supplied by central boiler plant. The seven condenser packs are currently air cooled. It is proposed that small plate heat exchangers would be installed to recover heat into the water loop.

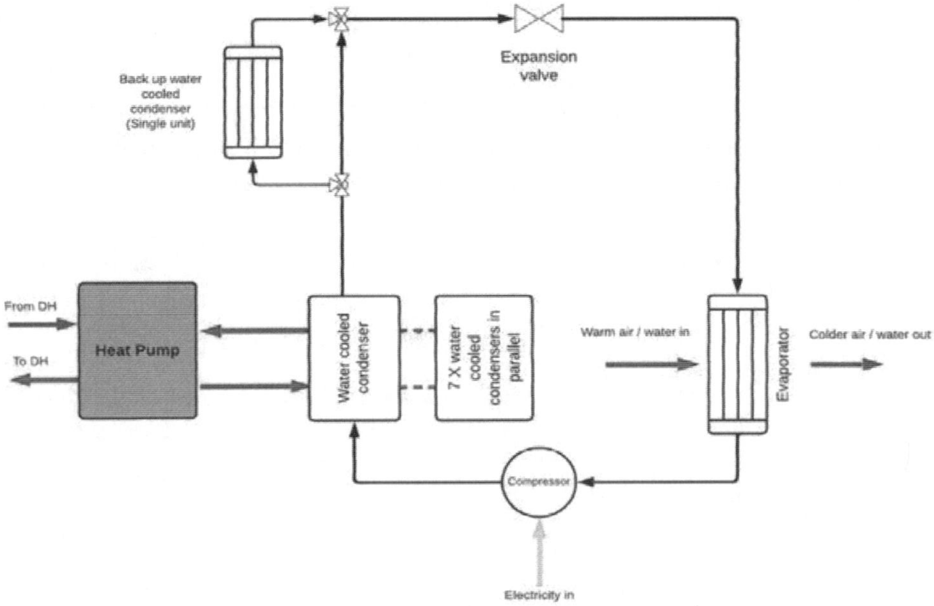


**Figure 22 - Configuration of the refrigeration system connection to the ambient loop**

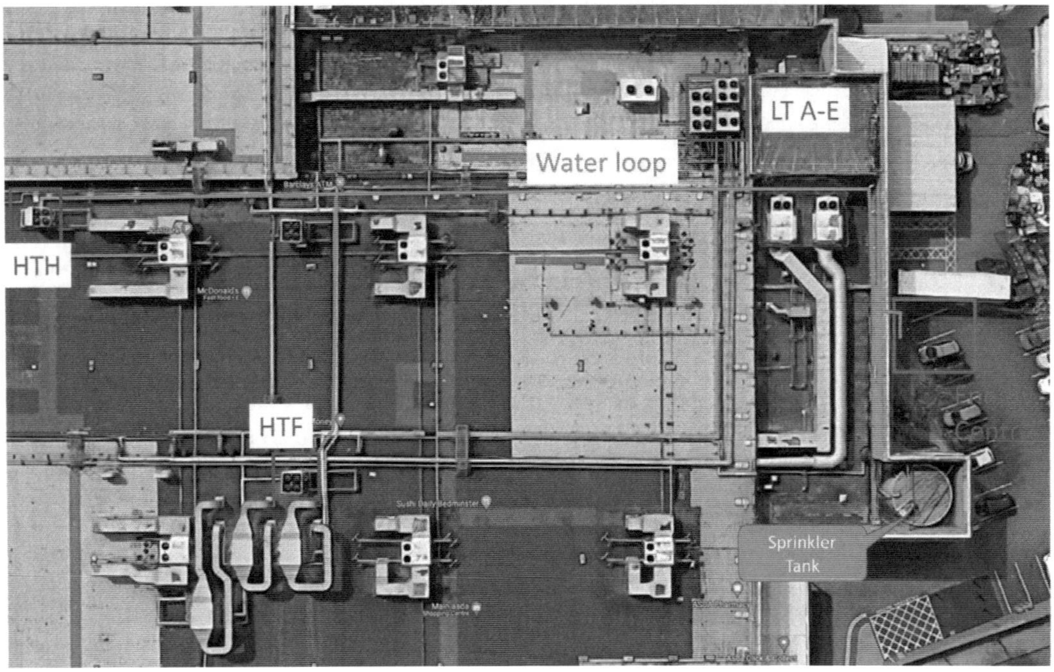


**Figure 23 - Connection of the refrigeration system connection to the ambient loop**

It is proposed that an existing large sprinkler tank could be used as a low temperature thermal store to balance waste heat recovery and heat pump operation. However, this is not viewed as an essential design element in replicating this approach.

The supermarket car park offers significant opportunity for installing EV charging. The charging options were assessed across a variety of use cases. They were centred around the needs of the delivery vehicles as they increasingly convert to EV, but also explored opportunities to maximise utilisation of the asset across customer, staff, residential and wider community charging opportunities.

The existing supermarket flat roof area is partially occupied by air handling units and refrigeration packs. There is scope to install solar PV around the plant which can provide electricity behind the meter and therefore avoid the tariffs applied to grid electricity. It is therefore proposed that 640 kW of PV is added on the roof. This is based on a roof area of 7,650 $m^2$ with a density of 6 $kW.m^{-2}$ when installed at a 10 degree angle with the required panel spacing. It has been assumed that only 50% of the roof would be available for PV given existing plant and shading from that plant.

Techno economic modelling has been carried out to establish economic and carbonomic viability. The concept design outlined above was developed into a model in energyPRO software and the CAPEX was investigated. The results showed that by integrating EV/ PV and heat it's possible to deliver carbon savings of 519 $Tonnes.a^{-1}$ and achieve a IRR of nearly 5%. More information on this study and its results can be found in Jones et al. (2023).

## 5. Wider Opportunities for recovery and reuse of Waste Heat

The capture and transport of waste heat represents a great opportunity for the decarbonisation of heat supply in buildings. To date, mostly high temperature waste heat has been reused and reported. However, with the recent advent of low and ambient temperature (4th and 5th generation) district energy networks, there is scope for the recovery and utilisation of heat from a range of novel, low temperature sources. The paper thus far has focussed on the opportunities for cooling and heat recovery with case studies from a supermarket, London Underground and a data centre. However, we are interested in the total potential for these and other applications. This section summarises some analysis we carried out to understand the potential.

A study has been undertaken which explored less conventional heat sources than had been investigated and compared on a national scale previously, namely data centres, electrical substations, wastewater, mine water, supermarket refrigeration, cold stores and underground railways. All of the heat output from these sources was at temperatures between ambient and 100°C, although with some variation in temperature for each source, both daily and seasonally. Data reported included: (a) the source of waste heat; (b) temperature(s); (c) number of heat sources; (d) method of calculation of heat source size; (e) total annual thermal energy output (for the selected heat source type); (f) breakdown of the percentage of annual heat output that is available within local authorities classified as predominantly urban (DEFRA, 2016). Source-specific methodologies were employed to estimate the different heat source quantities.

A previous study investigated waste heat opportunities in Scotland (Sinclair and Unkaya, 2020), therefore, the current study focused on the availability of a range of low temperature waste heat sources in the UK, excluding Scotland, but covering England, Wales and Northern Ireland (EWNI).

The low temperature waste heat application with greatest potential for reducing carbon emissions is for space or hot water heating in buildings, which are alone responsible for 23% of UK emissions (BEIS, 2021). A number of other uses for low temperature waste heat have been suggested e.g. adsorption cooling, separation and purification using membrane distillation and adsorption desalination, and electricity generation, for example using reverse electrodialysis (Luberti et al. 2022). However, efficiencies for these processes are quite low, so the main focus of this study was on delivering space heating and hot water i.e. contributing to meeting heat demand.

To qualify for financial support for the development of new and existing low carbon heat networks in the UK, a minimum heat demand of 2 $GWh.a^{-1}$ must be forecast (BEIS, 2022). Based on this requirement, for the current study a threshold single heat source size requirement of 250 kW and above was applied for all of the waste heat sources considered.

A summary of the source specific methodologies applied and the data sources utilised for each of the investigated heat sources is provided in Table 4.

For the individual low temperature heat sources, where applicable, the annual thermal energy output was divided in groups covering specific percentiles (<20$^{th}$, 20$^{th}$ – 40$^{th}$, 40$^{th}$ – 60$^{th}$, 60$^{th}$ – 80$^{th}$, >80$^{th}$), in order to represent the distribution of waste heat between sites of different sizes. An example of the results for estimated thermal output for one of the heat sources, namely data centres, is shown in Table 5 and Figure 24.

The majority of data centres cool the IT equipment using air, with waste heat in the form of heated air, exiting the server racks at 30-40°C. This heat could be harvested, for example by using an air-to-water heat exchanger, upgraded if required using a heat pump, and then distributed to users via a network. However, this would require significant modification of the data centre infrastructure, so is probably only viable for new build data centres. An alternative approach is to harvest the heat at 10-20°C from the chilled water heat rejection line. This would provide lower temperature heat but would be a far less invasive method as the connections would typically be located outside the building, so would be feasible for existing data centres.

Table 4 – Summary of methodologies and data sources utilised for each investigated heat source.

| Category | Heat Source | Waste heat calculation methodology | Initial size of data set | Size of data set after threshold applied | Summary of data sources |
|---|---|---|---|---|---|
| Electrical systems | Data centres | Potential determined from declared IT load or estimated from reported white space areas and typical PUE values. | 261 | 261 | Publicly available data sheets for individual data centres compiled together, including from manufacturers (Schneider Electric, 2019) and sector-specific portals (Colox, 2021, Data Centre Catalog, 2021) |
| | Electrical substation transformers | Potential determined based on assumed loading factors and heat loss factors from transformer nameplates. | 5794 | 1336 | Substation fleet data were obtained from personal communication and online network connection maps, e.g. (Northern Powergrid, 2020, SSEN, 2020). Data were obtained for each of the 7 DNOs as well as the TNO operating in the UK. |
| Water resources | Wastewater treatment plants | Potential determined by converting load entering values, in population equivalent, to volumetric flow rates, and then applying a heat extraction rate. | 1674 | 985 | Data on entering load for all WWTPs in the UK above 2,000 PE, based on the EU's Urban Wastewater Treatment Directive [European Commission 2017], and heat extraction rates obtained from literature (Cipolla & Maglionico, 2014, Dénarié et al. 2021, Neugebauer et al. 2021) |
| | Mine water | Potential determined based a pro-rata approach relating number of mines with total area per country, and then applying an average heat extraction rate. | 18584 | 18584 | Data on number of mines obtained from (Gluyas, 2020], whilst coverage areas per country were obtained from the literature [Johnston, 2008, Merrill & Kitson, 2017, Gillespie et al, 2013). Average heat extraction rate estimated from an existing UK case study (Evans, 2020). |
| Refrigeration systems | Supermarkets | Potential determined by applying typical COSP and refrigeration energy consumption factors to store annual energy consumption, which was obtained based on floor area data. | 4853 | 1424 | Typical specific energy consumption values, refrigeration energy use factor and COSP obtained from the literature (Tassou et al, 2011, Foster et al, 2018a, Foster et al, 2018b, Foster et al, 2019), whilst floor area data for England and Wales obtained from the Valuation Office Agency. |
| | Cold stores | Potential determined by applying typical COSP and specific volumetric energy consumption values to surveyed data on cold store volumes. | 241 | 193 | Typical COSP, specific energy consumption and cold store volumes obtained from a previous study carried out by one of the authors from this study [Evans et al, 2014]. |
| Underground railways | Ventilation shafts | Potential determined based on heat recovery rates from case studies, which were extrapolated based on vent shaft flow rate data from railway operators. | 65 | 65 | Heat extraction rates obtained from previous case studies (Davies et al, 2019, Lagoeiro et al, 2019), while flow rate and ventilation shaft location data obtained from a feasibility report by Transport for London (TfL, 2020) and personal communication with Tyne and Wear Metro (NEXUS, 2020). |

| Country | Number of sites > 250 kW | Share of waste heat in urban areas | Waste heat output (MW) |
|---|---|---|---|
| England | 250 | 94% | 1813 |
| Wales | 4 | 100% | 85 |
| Northern Ireland | 7 | 58% | 23 |
| Total | 261 | 93% | 1921 |

**Table 5 – Estimated waste heat from data centres and the shares located in urban areas**

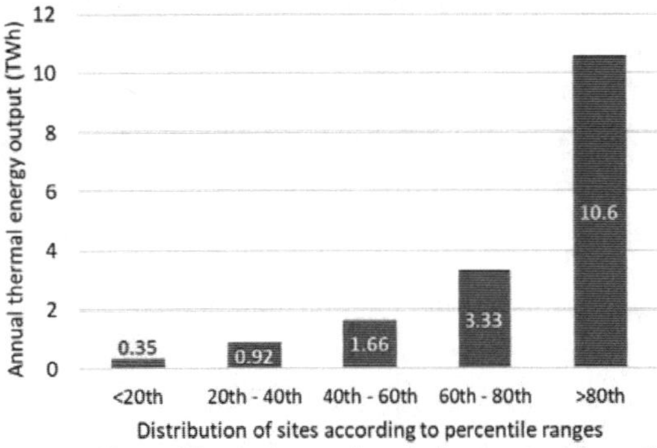


**Figure 24– total annual waste heat output from the data centres for different site size percentile ranges.**

It is seen from Table 5 that 93% of this waste heat is being generated in urban areas, while Figure 24 shows that 63% of the total potential is coming from the 20% largest sites. The total heat output for data centres was estimated to be 16.8 TWh.a$^{-1}$.

Similar analyses were carried out for all of the other heat sources. The results for estimated total annual thermal energy output for all sites identified for EWNI are shown in Figure 25.

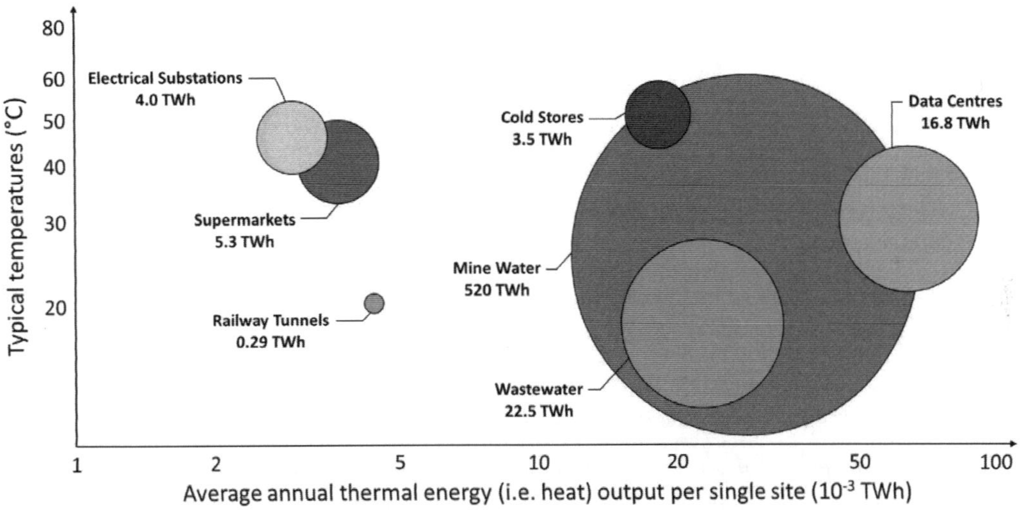


**Figure 25 Summary of total annual thermal energy outputs for heat sources considered**

The total quantity of thermal energy identified from the low temperature waste heat sources considered in this study, for EWNI, was estimated to be 572 TWh.a$^{-1}$, which would represent 132% of the total energy consumption for heat in these countries (432 TWh.a$^{-1}$).

There is an urgent need to reduce dependence on fossil fuel based energy sources for heating and to minimise carbon emissions, if the UK's target of zero greenhouse gas emissions by 2050 is to be met. Low temperature waste heat offers the potential of a very large heat resource, which has not been widely explored to date.

## 6. Conclusions and Future opportunities

If the UK is going to reach net zero by 2050 then it will need to address the hard to abate areas and this includes heating and cooling. This paper describes the research journey from sustainable cooling, to bivalent heating and cooling, waste heat recovery to smart local energy systems. Cooling and heating are a critical part of the energy system and there is significant opportunity to exploit this potential in decarbonising urban areas.

Cooling of London Underground using groundwater is achievable with a COSP of around 20, but barriers to implementation are capital cost and maintenance, particularly particulates. The PAHU study found that a wide fin space facilitates self cleaning and this has been tested and proven in a number of studies. By combining heating and cooling as described in Bunhill and with MICAH economic viability improves. This is further improved by combining into the who energy system as the GreenSCIES concept where heat recovery is central to viability. A number of opportunities for capturing waste heat in the UK are described.

The paper provides a comprehensive review of many of the research studies carried out over the last 25 years at LSBU with contributions from many.

## 7. References

Ampofo, F, Maidment, G., Missenden, J. (2003) Underground railway environment in the UK Part 2: Investigation of heat load, https://doi.org/10.1016/j.applthermaleng.2003.10.018 Applied Thermal Engineering.

BEIS (2017) Prices of fuels purchased by non-domestic consumers in the United Kingdom. Online: https://www.gov.uk/government/statistical-data-sets/gas-and-electricity-prices-in-the-non-domestic-sector accessed: 7 July 2017

Celsius (2020) https://celsiuscity.eu/heat-recovery-from-the-london-underground-in-islington-united-kingdom/ accessed 17 February 2023

Cipolla, S. & Maglionico, M. (2014) Heat recovery from urban wastewater: Analysis of the variability of flow rate and temperature. Energy and Buildings 69, p. 122-30.

Colo-X (2021) UK data centre database. Colocation Exchange Limited., Copyright 2016–2021. Available online: https://www.colo-x.com/data-centre-database/uk/ accessed 12 October 2021

Davies, G., Boot-Handford, N., Curry, D., Dennis, W., Ajileye, A., Revesz, A., and Maidment, G. (2019) Combining cooling of underground railways with heat recovery and reuse. Sustainable Cities and Society 45 pp 543-552. Online: https://doi.org/10.1016/j.scs.2018.11.045 accessed: 17 February 2023.

Department for Business, Energy & Industrial Strategy (2018) Heat Networks Investment Project, Introduction to the Scheme. Available online: https://www.gov.uk/government/publications/heat-networks-investment-project-hnip-scheme-overview accessed 21 November 2018.

Department for Business, Energy & Industrial Strategy (2020) Digest of UK Energy Statistics 2020. Available online: https://www.gov.uk/government/statistics/digest-of-uk-energy-statistics-dukes-2020 accessed 18 August 2020

Department for Business, Energy & Industrial Strategy (2021a) Net Zero Strategy: Build Back Greener. Available online: https://www.gov.uk/government/publications/net-zero-strategy accessed 4 November 2021.

Department for Business, Energy & Industrial Strategy (BEIS) (2022) Green Heat Network Development Fund Round 1: Guidance for Applicants, March 2022, Available online: https://www.gov.uk/government/publications/green-heat-network-fund-ghnf accessed 15 November 2022

Department for Business, Energy & Industrial Strategy (2021b) Heat and Buildings Strategy. Available online: https://www.gov.uk/government/publications/heat-and-buildings-strategy accessed 26 October 2021.

DEFRA (2016) UK Government GHG Conversion Factors for Company Reporting Department for Environment, Food and Rural Affairs. Online: https://www.gov.uk/government/publications/greenhouse-gas-reporting-conversion-factors-2016 accessed 7 July 2017

DEFRA (2016) 2011 Rural-Urban Classification of Local Authority Districts in England: Methodology. Department for Environment, Food & Rural Affairs. Available online: https://www.gov.uk/government/statistics/2011-rural-urban-classification-of-local-authority-and-other-higher-level-geographies-for-statistical-purposes accessed 15 November 2022

Dénarié, A., Fattori, F., Spirito, G., Macchi, S., Cirillo, V. F., Motta, M. & Persson, U. (2021) Assessment of waste and renewable heat recovery in DH through GIS mapping: the national potential in Italy. Smart Energy 1, February 2021, (100008)

Energy Systems Catapult (ESC) (2020). Decarbonisation of Heat: Why It Needs Innovation. Available online: https://es.catapult.org.uk/brochures/decarbonisation-heat/ accessed 23 June 2021.

European Commission (2017). Urban Wastewater Treatment Directive (UWWTD) site for the United Kingdom. Available online: https://uwwtd.eu/United-Kingdom/ accessed 15 November 2022.

Evans, J,A. Hammond, E.C., Gigiel, A.J., Reinholdt, L., Fikiin, K. and Zilio, C. (2014) Assessment of methods to reduce the energy consumption of food cold stores. Applied Thermal Engineering 62 pp 697-705

Evans, L. (2020) South Seaham Garden Village District Heat Network Available online: https://www2.groundstability.com/wp-content/uploads/2020/02/04-Sustainable-Energy-SGV-District-heating-opportunity.pdf accessed 15 November 2022

Evening Standard (2003) West End Final, Tuesday 15 July 2003, London.

Evening Standard (2023) London commuters struggle in heatwave on Tube https://www.standard.co.uk/news/london/london-commuters-struggle-temperatures-tube-central-district-bakerloo-line-b1013140.html accessed 18 July 2022

Foster, A., Evans, J. and Maidment, G. (2018a) Benchmarking of supermarket energy consumption. 5th IIR Conference on Sustainability and the Cold Chain, Beijing, China, 2018.

Foster, A., Hammond, E., Brown, T., Maidment, G. and Evans, J. (2018b,) Technological Options for Retail Refrigeration. International Institute of Refrigeration. Paris, France, 2018. ISBN: 978-2-36215-029-6, ISBN: 978-2-36215-032-6

Foster, A., Brown, T., Evans, J. and Maidment, G. (2019) Benchmarking of supermarket energy consumption. 25th IIR International Congress of Refrigeration, August 24-30, 2019, Montreal, Canada.

Gluyas, J.G., Adams, C.A. and Wilson, I.A.G. (2020) The theoretical potential for large-scale underground thermal energy storage (UTES) within the UK Energy Reports 6 pp 229-237

Gilbey M. Thompson, J. (2009) TEMPERATURE CONTROL ON LONDON UNDERGROUND SME Annual Meeting Feb. 22-Feb. 25, 2009, Denver,

Gillespie, M.R., Crane, E.J. and Barron, H.F. (2013) Study into the Potential for Deep Geothermal Energy in Scotland: volume 2 Scottish Government Project Number: AEC/001/11 British Geological Survey Commissioned Report, CR/12/131. Available online: https://www.gov.scot/publications/study-potential-deep-geothermal-energy-scotland-volume-2/ accessed 15 November 2022

Innovate UK (2019) Prospering from the energy revolution https://www.ukri.org/what-we-offer/browse-our-areas-of-investment-and-support/prospering-from-the-energy-revolution/ accessed 20 February 2023

Johnston, D., Potter, H., Jones, C., Rolley, S., Watson, I. and Pritchard, J. (2008) Abandoned mines and the water environment - Science project SC030136-41 Publ. by The Environment Agency. Available online: https://assets.publishing.service.gov.uk/government/uploads/system/uploads/attachment_data/file/291482/LIT_8879_df7d5c.pdf accessed 15 November 2022

Jones P., Dunham, C, Fenner, R., Rozinski, K., Marques, C, Maidment G. (2023) Energy Superhubs - the use of supermarkets as local energy centres, Submitted to CIBSE Technical Symposium 2023

Lagoeiro, H., Revesz, A., Davies, G., Maidment, G., Curry, D., Faulks, G. & Bielicki, J. (2019) Heat from Underground Energy London. Proceedings of the 25th IIR International Congress of Refrigeration, 24-30 August, Montréal, Canada.

Lagoeiro, H., Revesz, A., Davies, G., Gysin, K., Curry, D., Faulks, G., Murphy, D., Vivian, J. & Maidment, G. (2021) Waste Heat Recovery from Underground Railways – Evaluating the Cooling Potential. Proceedings of the CIBSE Technical Symposium, UK, 13 – 14 July, 2021.

Lagoeiro, H., Wegner, M., Revesz, A., Davies, G., Curry, D., Vivian, J., Faulks, G., Murphy, D. & Maidment, G. (2022) Recovering Waste Heat from the London Underground: Sizing the Opportunity WSED Young Energy Researchers Conference, Wels, Austria, 2-3 March 2022

Luberti, M., Gowans, R., Finn, P. and Santori, G. (2022) An estimate of the ultralow waste heat available in the European Union Energy 238 121967

Merrill, T. and Kitson, L. (2017) The End of Coal Mining in South Wales: Lessons learned from industrial transformation Global Subsidies Initiative (GSI) Report Publ. by The International Institute for Sustainable Development Available online: https://www.iisd.org/system/files/publications/end-of-coal-mining-south-wales-lessons-learned.pdf accessed 15 November 2022

Neugebauer, G., Kretschmer, F., Kollmann, R., Narodoslawsky, M., Ertl, T., Stoeglehner, G. (2015) Mapping Thermal Energy Resource Potentials from Wastewater Treatment Plants. Sustainability 7, 12988-13010.

NEXUS (2020) Transport for Tyne and Wear (2020) Personal Communication.

Northern Powergrid (2020) Generation Heat Map. Copyright Northern Powergrid Holdings Company 2020 Available online: https://www.northernpowergrid.com/generation-availability-map accessed 6 December 2020

Ofgem (2016) Tariffs and payments: Non-Domestic RHI.

Revesz, A. Maidment, G Marques, C Davies, G Jones, P Dunham, C Tozer, R Matabuena, R Low, G Hamilton, V Kelly, D Hampton, C Scott, J, Bond, C Murawa, M Talbot, D Sahotay, G., Curry D, (2019) GreenScies - Green Smart Community Integrated Energy Systems. 5th Int Conf on Smart Energy Systems. Copenhagen. https://smartenergysystems.eu/wp-content/uploads/2019/09/3-3_AkosReveszSESAAU2019.pdf

Sinclair, C. and Unkaya, G. (2020) Potential sources of waste heat for heat networks in Scotland. Available online: https://era.ed.ac.uk/handle/1842/37445 accessed 15 November 2022

Skidmore, C. (2023) MISSION ZERO - Independent Review of Net Zero, MP https://assets.publishing.service.gov.uk/government/uploads/system/uploads/attachment_data/file/1128689/mission-zero-independent-review.pdf accessed February 2023

Schneider Electric (SE) (2019) Data Center Reference Design Selector. Available online: https://www.se.com/us/en/work/solutions/for-business/data-centers-and-networks/reference-designs/ accessed 15 November 2022

Scottish and Southern Electricity Networks (SSEN) (2020) Generation Availability. Copyright 2020 Scottish and Southern Electricity Networks Available online: https://network-maps.ssen.co.uk/opendataportal accessed 5 December 2020

Tassou, S. A., Ge, Y., Hadawey, A. and Marriott, D. (2011) Energy consumption and conservation in food retailing. Applied Thermal Engineering 31.2-3: pp 147-156

The London Paper (2009) TfL's rush to cool Tube, Monday 29 June 2009, London.

Thompson JA, Maidment GG, Missenden JF. (2006) Modelling low-energy cooling strategies for underground railways. Appl Energy 2006; 83: 1152–1162.

TfL (2020) Early Market Engagement - TfL Waste Heat Opportunities. Available online: https://procontract.due-north.com/Advert?advertId=91605c8a-5d89-ea11-80ff-005056b64545 accessed 15 November 2022

TFL (2021) London Underground Average Monthly Temperatures, https://data.london.gov.uk/dataset/london-underground-average-monthly-temperatures, accessed 20 February 2023

TfL (2022) TfL trials innovative cooling solution designed to reduce temperatures on the Tube network https://tfl.gov.uk/info-for/media/press-releases/2022/july/tfl-trials-innovative-cooling-solution-designed-to-reduce-temperatures-on-the-tube-network accessed 20 February 2023

UITP (2015)

UK Government (2021) Plans unveiled to decarbonise UK power system by 2035, https://www.gov.uk/government/news/plans-unveiled-to-decarbonise-uk-power-system-by-2035 accessed 20 February 2023

## Acknowledgements

This paper is presented in response to the award of the J&E Hall Gold medal and the recognition of this is very much valued and appreciated. J&E Hall introduced me to cooling (and heat pumps) 40 years ago as a technical apprentice. They provided me with a fantastic foundation to a brilliant industry. Thank you to colleagues at JEH – Andrew Bowden, Mark Roberts, Tony Hirschfeld, Peter Dalton, the late Dave Ball, Gordon Flaherty, Terry Young and JEH friends.

This paper also acknowledges the team work of many colleagues over many years including:-

- Colleagues from LSBU including Professor John Missenden, Professor Ron James, Dr Jolyon Thompson, Dr Felix Ampofo, Dr Catarina Marques, Dr Henrique Lagoeiro, Dr Gareth Davies, Dr Helen Turnell, Aya Heggy, Dr Alex Paurine, Dr Akos Revesz and Matt Wegner.
- Colleagues from London Underground Nick Boot-Handford, Dan Curry and Ernie Boddington from London Underground
- The GreenSCIES consortium including Professor Phil Jones, Chris Dunham, Kristina Rozinski and Russell Fenner.
- Also Joel Hamilton from BEIS.

This paper acknowledges the support and funding from Innovate UK, EPSRC, London Underground and the EU. It is also dedicated to our wonderful colleague Carole Bond who was an inspiration to those who knew her.

## About the author

Professor Graeme Maidment FinstR received the J&E Hall Gold Medal in recognition of his practical achievement in demonstrating innovation in delivering low-carbon solutions for the refrigeration, air conditioning and heat pump sector.

Graeme has pioneered the integration of heating and cooling with practical projects on combined heating, cooling and energy storage; integration of heat, power and whole energy systems; and smart heating and cooling. He has also led several initiatives to promote sustainability in our sector such as refrigerant leakage reduction.

He is currently working on low carbon strategies in the food sector and supporting cold chains in developing countries, is the lead advisor on Heating and Cooling for BEIS the Dept for Business, Energy and Industrial Strategy and co-lead of the Innovation Challenge "Affordable Heating and Cooling of Buildings" where he works with 24 countries on clean energy solutions.

## DISCUSSION REPORT

**Dermot Cotter**, chairing, thanked the author for a very interesting paper

**Dermot Cotter** – Air cooling coils for the London Underground. Isn't the London Underground that dirty that they get contaminated more than most places? Is it a challenging environment down there.
**Graeme Maidment** – I was expecting that question. We're engineers and what we do is provide solutions for engineering challenges. We worked with what we had and we didn't particularly examine the particulates. I know that there are people at Imperial College that have done this and if anyone is interested, I would refer them to that. Particulates are pretty much everywhere but it is something that we need to try find ways to avoid. We need to work with much cleaner systems and that's not just on the tube. We need to work cleaner in cities. If you look at air quality in cities in general, it's not very good. A lot of it is cited as being due to road transport, I don't necessarily agree. I also think boilers and the use of fossil fuels are things that we need to avoid as well.

**Dermot Cotter** – Why do you think the dirt itself doesn't stay on the coils. Is it just less friction?
**Graeme Maidment** – We observed them and there's something called air washing, which I think is used on stoves and things like that, where a level of particulates gets to a certain level where the forces, the velocities are sufficient to drag them away. The problem on the narrow fins was that the build-up of the particulates didn't get to sufficient velocity to drag them away before the coil was constrained. Once you narrow the gap the velocity increases and the forces were sufficient to drag it away. We haven't done any computation or any physics analysis. That was from just observation.

**Dermot Cotter** – That sounds very practical. Sometimes science doesn't tell you the answers and practicality does. Do you have any economic details available for the likes of the green scies project. What about the investment, the capital, and should there be more investment like this?
**Graeme Maidment** – I think it's really important that these things are going to be driven by cost. The scheme that we presented, the capital expenditure is about £16 million to construct and the reduction in operating expenditure is about £600,000 or £700,000. So the pay back is about 12, 13, 14 years. It does pay back, the pay back is slow, but the comparison is with gas boilers. I think at some point we are going to have to move away from comparing against gas boilers. We need to compare against alternatives and if the alternatives are going to be Hydrogen or air source heat pumps then we really stack up well. Commercially it makes sense and compared to air source heat pumps and Hydrogen it makes lots of sense. Practically as well in some of these areas, if we work on the basis that we're not going to be able to use gas boilers, which if we're going to get to net zero without Hydrogen, we're very unlikely to, then it's where are you going to put equipment as well and having lots of individual heat pumps everywhere is probably not going to work.

**Dermot Cotter** – Have TfL looked at designing their braking systems differently to reduce this heat or is it because they are slowing down the trains so much that you have the friction and that there is no way of preventing it? Or is there anything else they can do to reclaim this heat i.e. by not having it at all?
**Graeme Maidment** – You are completely right and it would be good to avoid generation of heat in the first place and in generating the heat you are introducing a loss. On the Jubilee Line coming into stations, you go uphill and when you come out of the station you go downhill. That is probably the simplest way of benefitting from potential energy. There is something called regenerative braking, and I'm not a transportation engineer so I don't know about trains. Although we did do some work on fitting air conditioning for trains and it's very, very complicated fitting air conditioning to underground trains, primarily because it's all controlled by the safety case. The safety case is very stringent but there are also so many other factors involved, civil engineering, looking at the rolling

stock and whether the power is sufficient, looking at clearance. It's actually very complicated to do anything on an existing train so I believe that the cost of introducing regenerative braking would potentially be prohibitive.

**Dermot Cotter** – Now we are going down the route of getting rid of gas boilers and moving to electricity is there a concern about having a single source of energy. What happens if that fails?
**Graeme Maidment** – That is a really good question. Energy security, resilience and balancing is something that has to evolve over time and before we talk about resilience we need to talk about balancing. If we are going from electricity being used for predominantly small power and some industrial and commercial cooling to small power plus heat pumps plus cooling plus EV then there is going to be a big shift in the demand. The projections are that it is going to double or triple. Balancing the supply, when there is renewable energy available, with the demand is a big need and there are a number of ways that will happen. The use of batteries, electric to grid batteries, the use of batteries in cars but also using demand flexibly will help. Whether this helps with the resilience aspect, it's bound to give some degree of resilience. It is a good question because we need to make sure we have energy security, we want to be in control of our own destiny but we need to make sure we are resilient in that as well.

**Dermot Cotter** – When you are dealing with projects like the glassworks is there any concern that you have put in a fantastic heat pump system and then 5 years down the road the glassworks closes and you have lost all of that heat? Has that been considered?
**Graeme Maidment** – Yes, that was one of the questions that was asked about the data centre. What's the long-term commitment of the data centre? We are in a changing world and things move really quickly economically, there are many new developments in terms of processes and how data centres work. Do we want them close to the community or do we want them outside? Sometimes people said put them in isolation. Now they are moving to be called edge data centres where they are in the locality. I think what we tried to do in GreenScies is to give some reassurance to the owner and the operator that whoever is the heat source has some level of commitment to being there for a period of time and normally these projects are run for 20, 30, 40 years. It was unrealistic to expect a glassworks or a data centre to sign up for that period of time but potentially 10 to 15 years is realistic. Then the risk has to be mitigated but over time there could be lots of these schemes all interconnected so you are not just reliant on one heat source, you could have multiple heat sources. It is definitely something that is considered, it is in the risk register, and it is something that we try and mitigate but it is still a residual risk.

**Dermot Cotter** - Where does the 600,000 heat pump figure come from and what do you think the refrigerant of the future will be?
**Graeme Maidment** –I don't know where the figure of 600,000 heat pumps by 2028 came from but I think it's a really good number. I don't know the figure we are at today, maybe 50,000 a year, so we have to move quickly. The important thing is to get into the market and we all want to do the right thing. Already the number of heat pumps being introduced and the lead time for heat pumps is increasing significantly. I think it's a good number, it's an ambitious number and if we can get to it then it really sets us up for the future.
The refrigerant of the future is a really interesting one and the challenge will be that we've got, potentially, F Gas Regulation changes happening and we will, perhaps, align with some of the changes that Europe introduces. The availability of higher GWP refrigerants is going to be at risk and therefore we are going to need to look at other alternatives. Also we have to consider the PFAS concerns that could put certain refrigerants under threat. We may need to look more carefully at Hydrocarbons for domestic and look at how we minimise charge and we minimise risk for them and for larger scale systems. If we can't find a HFC refrigerant that has a long term future then we need to rethink about Ammonia and Hydrocarbons.

**Dermot Cotter** – They must be fit for the location and application and they must be value for money.
**Graeme Maidment** – Don't forget that in the centre of London there is an Ammonia heat pump fitted and it works.

**Dermot Cotter** – What is the economics of moving hot water around? Is it viable?
**Graeme Maidment** – It makes a lot of sense because the temperature of the water that moves around is ambient temperature so it's 15 to 20° C and you can vary it if you want and by varying it you can give some storage within the pipework itself. The pipework heat losses are negligible to the extent that we did some analysis and it's not even worth insulating them. The pumping costs are relatively small. The nice thing about moving it around is that you can move it to where there is a cold source, you can take that cold source and you can move that cold bit along and you can easily move it to a heat source and move that heat so a 5th generation, an ambient loop can potentially keep going and keep going and keep going and as long as you've got good balance between heat and cold it makes so much sense. The losses are not a big issue, but the benefit is much greater of bivalently heating and cooling.

**Dermot Cotter** – Thank you Graeme for an interesting paper and for answering these questions.

IOR.org.uk
REFRIGERATION AIR CONDITIONING HEAT PUMPS

The knowledge hub for refrigeration, air conditioning and heat pumps

# Is Money Best Spent on Efficient Design or Service in Refrigeration

Robert Unsworth FInstR

Copyright © 2022
Institute of Refrigeration

**Why you should attend:**

- To get an overview of the key design and maintenance activities that are critical to achieving good system performance.
- To hear about the return on investment for both capital expenditure (CAPEX) and for service and maintenance activities and how the two are related.
- To hear about a holistic view of system performance and how this can be maintained and improved.

## Abstract

In 2010 approximately 6.3 billion tonnes of food were produced globally, it is estimated that 2.3 billion tonnes of this require a refrigeration process of some sort. This has been increasing by approximately 10% year on year.

Refrigeration systems are responsible for consuming approximately 17% of all power generated globally. Assuming an average coefficient of performance (COP) of 4, we as an Industry are using around 4,500 billion kWh per annum and rejecting roughly 18,000 billion kWh of heat. Carbon emissions attributed to the operation of refrigeration systems vary depending on the energy source the system operator chooses to employ. As a sector, therefore, we are producing rather a lot of carbon, just in running our refrigeration systems. Arguably, not all will be efficient and from experience, most are not. Any decrease in efficiency just means greater amounts of cost, wear, and carbon emissions (ignoring altogether refrigerant type and leakages, which is a whole other topic of discussion, although related).

With the enormous increase in the adoption of heat pumps (a chiller working at a higher pressure) expected in the coming years, many of which will be attached to existing and new refrigeration systems to boost the Thermal Heat of Rejection up to a useful temperature for application in another process, it would be nonsensical to spend capital investment on an efficient solution without considering how effective your heat source is first, or would it?

Many industrial refrigeration systems have been in operation, and are expected to be so, for multiple decades, and in fact, I regularly come across compressors that are older than I. If one investigates the life cycle cost of these systems, the largest portion of the expenditure is energy, outweighing all other costs by a factor of 5 to 8.

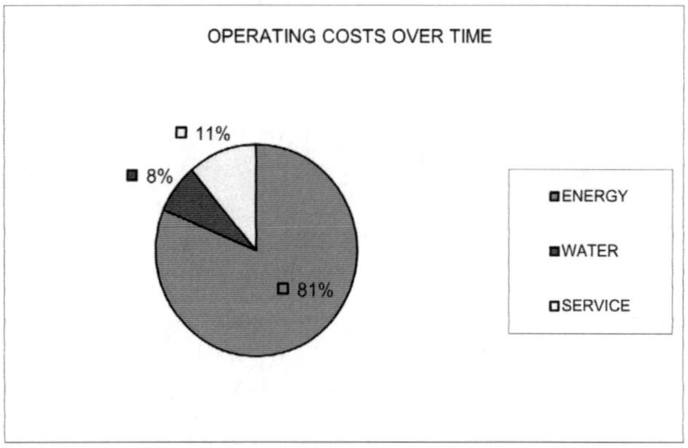


**Figure 1: 1MW -22OC Freezer Store Operational Cost breakdown**

When using an evaporative system to reject the heat from the refrigeration process, the greater the efficiency, the lesser the water consumption and effluent demand. With water availability being put under more and more pressure globally, we are now placing a strong focus on reducing system demand at the design stage; whereas historically, it was generally an afterthought, and arguably an inefficient refrigeration system uses more than it needs to.

Inefficiencies in refrigerated spaces, whether they be lights/computers left on, conveyor friction, motor efficiencies on drive lines, and of course staff running around, end up increasing the heat load and therefore increasing the demand on the refrigeration system. These inefficiencies, when combined with an inefficient refrigeration system all add up to the energy wasted from the system which is rejected as heat at the condenser.

The following paper attempts to explain a little more as to why designing efficient refrigeration systems is critical in solving many issues related to carbon, water, and energy reduction along with other additional costs which impact service expenses.

The argument: Is Money Best Spent on Efficient Design or Service in Refrigeration?

In my opinion, there is little point in increasing your CAPEX to build the world's most efficient refrigeration system, only to forget about it once commissioned and not maintain it in the appropriate manner. It will not be efficient or reliable for long.

## 1. Introduction

Being a bit of a geek, my first thought when being asked to write this paper and take part in the discussion arguing from a technical perspective as to why spending funds on efficient refrigeration design is more important than funds spent on service was that it should be relatively easy when looking into a life cycle cost calculation. Of course, I found myself arguing with myself because of course, both are of equal importance.

There is little point on increasing your CAPEX to drive efficiency up in a refrigeration system, which would arguably lead to larger/more efficient heat exchangers, larger slower running variable speed compressors and the correct type of condenser if they are not maintained or cleaned. For example, 1mm of calcium deposit on an evaporative condenser will easily eat up the cost of an entire year's Service Contract on a large installation.

Similarly, oil logging inside evaporators, or ice build-up on external heat exchange surfaces will equally have a negative effect. Pumps, circulation rates and the correct operation of the compressor itself, all have roles to play in system efficiency, but none will run optimally if they are not well maintained, even when the end user operates the system as originally designed and intended (often not the case).

Depending on the type of system, the arguments change somewhat, but all depend upon a simple fact. A refrigeration system is a closed loop (or at least it should be). Any energy entering the system will also leave the system, one way or another. As the second law of thermodynamics breaks down heat travels toward cold, energy will enter the system on the cold/low-pressure side of the system (intentionally or not) and it will leave (intentionally or not) if the temperature around it is either warmer or colder depending on which side you are looking at.

Heat we intend to enter the system on the cold side would be via evaporators and possibly intentional superheaters for example. Heat we do not perhaps intend to enter would be via poor insulation, or insulation breakdown leading to evaporation of more liquid or superheat of dry suction pipes. The first is controlled by efficient design and system control strategy, the other by proper maintenance.

To put some perspective on this point, if you have ice build-up on a suction pipe for example, it has twice the thermal conductivity of glass (0.25w/m.k) and half as much as cement (1-1.7w/m.k), but less than steel and copper (385w/m.k), obviously much less than insulation (eg. 0.035w/m.k). Therefore, if the vitally important vapour barrier breaks down on your pipework and you experience ice building up under and through your insulation, it is not a bad conductor of heat. Keeping the heat and moisture away from the places where it is not intended, is only achievable by a combination of efficient/proper design and proper maintenance.

To make some base calculations of efficiency on a system, I have chosen a refrigeration system operating with a suction condition of -30$^{o}$C and condensing at +32$^{o}$C, with a room temperature of -22$^{o}$C and an ambient wet bulb temperature of +21$^{o}$C and a load of 1,000kw. (Typical design considerations in the UK)

The system being considered is a pump circulated ammonia system with a pump overfeed of 3:1 and an evaporative condenser with a high-pressure receiver.

The application is something like a cold store, but to be honest this is irrelevant as I am considering the refrigeration cycle alone and not how often the clients' operators leave the doors open in the store for example, but I will consider the effect (snow/ice build-up) this has on our system.

Therefore, I will make some calculations and observations based on these conditions and the effect on running costs with different considerations and maintenance related issues.

Once all these are mapped, it builds up a stark picture as to where most of the end users' money is heading.

## 2. Evaporators

Evaporator design depends on the application, room shape, evaporator location, room type etc, but as this is where the energy enters the system and can be easily affected by external forces such as doors being left open, they are the first thing which tends to impact overall system efficiency.

For example, years ago it was standard to fit finned coils on ceilings and use natural convection to design evaporators. Due to the relatively poor heat transfer characteristics, a lot of surface area was required compared to the usual forced draught coolers we see today.

At the end of the day, if one needs a cold room, physics dictates that you need a colder surface to transfer the heat imposed on it.

Determining the required heat exchange surface involves many factors, but in essence, it is down to the logarithmic mean temperature difference (LMTD) and the heat exchanger's own thermal performance. Thermal performance is determined by many different factors, such as tube and fin design associated with air volumes and velocities across heat exchange surfaces, internal and external heat transfer rates, materials used, air conditions and lastly fouling.

If everything else is in place, the required heat exchanger surface area is calculated taking into consideration the design capacity and temperature difference.

The room temperature is fixed according to the user's requirements and evaporating temperature related directly to how large a surface area you wish to install. The closer the evaporating temperature is to the room temperature the larger the surface area of the heat exchanger will be. When evaporating temperature increases so does the suction pressure to the compressor. For every 1K you can raise the suction pressure, compressor performance increases by approximately 4%.

This 4% increase in compressor performance is seen as an increase in cooling capacity and Coefficient of Performance (COP). This is because the compressor has a certain swept volume, and as the saturated boiling temperature of the refrigerant increases, the specific volume of the vapour decreases, the density increases and the compressor's relative cooling capacity increases as it is moving a greater mass of refrigerant.

For example, 1,000kw of ammonia at -33°C boiled off, creates approximately 2,900m³/hr of vapour excluding flash gas (i.e. the compressor swept volume). The same 1,000kw of heat evaporating ammonia at -30°C would produce 2,550m³/hr of vapour. Therefore, having an evaporator which can operate in the higher evaporating condition would mean although you need a larger evaporator, you need a smaller compressor, and because it's operating at a higher evaporating pressure (1.19 bara rather than 1.03 bara) it does not have to compress the vapour as far to achieve the condensing pressure, therefore requiring less power per kW cooling. Less flash gas is also created at the expansion valve, which is a volume of vapour added to the evaporated vapour volume on the suction side of the compressor.

Reducing the liquid temperature before the expansion valve by subcooling reduces the amount of flash gas, and therefore frees up some of the suction capacity of the compressor for more "useful vapour" from an evaporator. This is one of the reasons for including things like economizers, compound compression and sub-coolers. Without this extra equipment, and in the example being considered, nearly a quarter of the swept volume of the compressor would be used just to deal with the flash gas.

Obviously, all these items cost money, but their effect on system efficiency can be substantial.

There is clearly a trade-off, as the evaporating temperature gets closer to the room temperature, you need proportionally much more surface area. And as the heat transfer figure reduces it becomes practically impossible and economically unviable to go larger.

Having said all that, once purchased and installed, these perfectly good heat exchangers/evaporators, economizers, and sub-coolers etc. are not going to perform very well and save you energy if they are covered in ice on the outside and coated in oil on the inside. Correct oil selection is an important factor to avoid fouling inside refrigeration systems.

For example, an evaporator with an air on temperature at -22°C and an evaporating temperature of -30°C with a circulation rate of 3:1, will need to evaporate close to -32°C to deliver the same capacity if it is covered with 2mm of ice. What tends to happen however, is that the air off temperature is not reached, because the suction condition remains the same in the compressor, and the room temperature raises a little, or the coolers have to run longer to compensate, as the same 2mm of ice has reduced the cooling capacity of the evaporator by nearly 16%.

Another issue which is not usually considered, is the fact that due to the over circulation rate of 3:1 being based on 50kw and fixed at commissioning, when the cooler is only delivering 42kw at best, the circulation rate has increased to 3.6:1, so in effect this additional over circulated liquid, ends up in the wet return, being sucked back through the risers and wet return pipework to the pump separator, increasing the pressure drop and worsening the evaporating pressure in the evaporator. A double impact, so to speak.

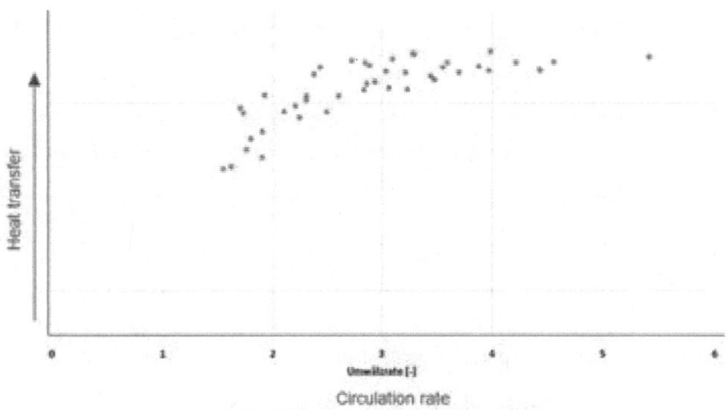


*Thanks to Thermofin for test information*

**Figure 2: Standard bottom fed pumped over feed rate thermal performance on evaporator (without distributor)**

Likewise, an oil film in the evaporator which increases a normal fouling figure to 0.0004 m2K/W would reduce the cooling capacity by 10%, so a combination of ice and oil really does hamper the efficiency of the system.

Although systems are hopefully designed to keep evaporators clear of ice and minimise oil contamination, it is widely acknowledged that it takes good service practices to ensure the system remains in optimal condition and operation throughout its lifetime.

Keep evaporators clean, in good working order and oil-free, and defrost them in a well-managed programme. When defrosting, keep the heat in the coil as much as possible, at cooler temperatures and do not allow defrost periods to extend past what is necessary. Remember, any heat and moisture entering the room from defrost must be removed again. However, if using hot gas to defrost an evaporator, your evaporator becomes a condenser and your condensing pressure will fall, so if controlled correctly, this can have a positive effect on your refrigeration compressor, and this will be covered a little later.

## 3. Compressors

As an employee of an international industrial compressor manufacturer, one might expect me to spend time discussing the pros and cons of various compressor technologies, but after far too many years on this subject I have decided to ignore the compressor type but focus on the system surrounding it, as I have often found people select compressor type for emotive reasons rather than fact, and so I would rather concentrate on items around the compressor which affect efficiency and service.

Starting on the suction side, suction strainers add pressure drops and should be clean. Similarly, on screw compressors and some manufacturers' reciprocating compressors these days, coalescent filters are used inside oil separators to reduce the oil carryover from the compressor package.

The coalescent filters are a compromise really by the compressor designer. Approximately 60% of the mass flow through a screw compressor is oil, therefore this needs to be removed. The compromise is, the better the effect of the filter, the greater the pressure drop and the greater the shaft power on the compressor. Over time, these pathways through the coalescers elements get blocked with oil deposits etc, and the velocity increases, causing a greater pressure drop and a higher oil carryover. This is where maintenance can come into play in either replacing or cleaning coalescers from time to time to keep the compressor running as it was intended.

There is also a return from the oil separator coalescer elements to the suction side of the compressor. This returns any oil which has been removed by the coalescer, back into the machine. This is normally controlled with a needle valve or a small orifice. Sometimes these lines can be blocked and therefore oil will collect at the coalescer and then get blown through and out of the machine into the system producing negative effects. These orifice/valves should be checked regularly but should not be opened too much as this is a direct return from the discharge to the suction side of the capacity and in some conditions could keep a compressor running irrespective if there is an actual cooling load on it or not.

Screw compressors, unless specifically ordered with variable Vi, compress to a certain pressure ratio. Therefore, if floating condenser control is used and the compressors' Vi position is set to summer conditions with larger differential pressures, the compressor will continue to compress to the higher pressure. As the gas expands as it leaves the discharge port, some of this energy is recovered, but not all. Auto Vi control moves the discharge position of the compressor automatically as the discharge pressure moves, keeping the correct compression ratio. The difference between having the correct ratio and not can be an excess in power consumption of up to 10%. Utilising Auto Vi in your system design represents a minor increase in CAPEX on a screw compressor application but is a good one to have, depending on operating conditions.

Some compressor manufacturers with fixed Vi have manually adjustable Vi slides which can be adjusted for example in the spring for summer running and in autumn for winter conditions. These adjustments should be carried out during a service visit. It is not unusual to come across a screw compressor which has been moved from a high-stage operation to a booster operation and the Vi position has been forgotten about.

Mechanical items such as motors, oil pumps etc should be in good working order. This goes without saying and is again a function of good maintenance.

Control and system design is also extremely important to ensure compressors are not cycling too often. For example, a variable speed reciprocating compressor running at a constant lower speed will run more efficiently, certainly last longer and cost less to maintain than if it were cycling up and down chasing load variations.

One also needs to bear in mind, especially with ammonia, that the "dry" suction line is full of saturated, un-superheated vapour. If the suction pressure raises because the refrigeration load has increased quickly, the compressor may not be following, and because there is a large thermal mass in the steel suction pipework as well as the suction side of the compressor itself, the vapour will condense on the insides of the "dry" pipework. This liquid will be sucked into the compressor, causing all sorts of issues. It's therefore important to control these capacity surges for both efficiency and equipment resilience reasons. A 3-5k suction pressure lift against 2 tonnes of steel suction pipework can condense a lot of liquid. In fact, it is about 5 litres at -30°C. This effect can be even higher in heat pump applications because of the higher suction gas density; therefore a controlled, stable load becomes even more important.

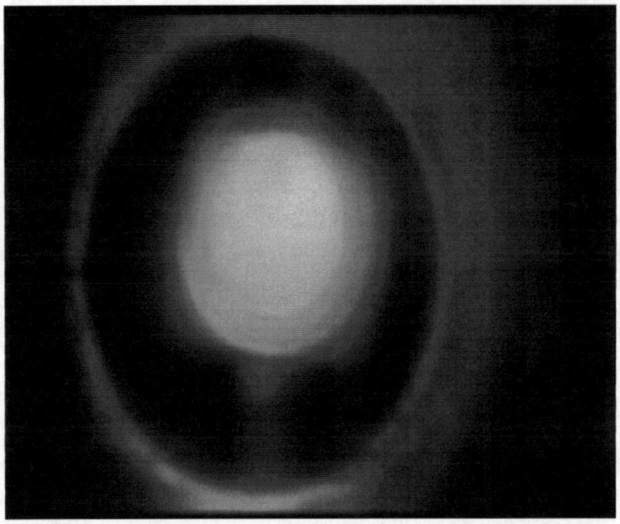

**Figure 3: This photo is the inside of a "dry" suction pipe where the suction pressure has risen by 3K. The lighter area and the "stream" along the bottom are the resultant condensed liquid being formed due to the rising suction pressure.**

## 4. Condensers

The choice of condenser type really depends on location and individual requirements such as local water and effluent costs against power, capacity, and location. However, in the UK these days it's more common than not that designers are using dry, adiabatic and hybrid condensers over traditional evaporative types on industrial systems.

In a similar way to evaporators, there is an optimal/cost-effective design point where the larger the condenser, the more fan power it will consume to achieve the unachievable as the condensing saturated condition reaches the dry or wet bulb temperatures.

The difference with the condenser is, we can design an efficient condensing surface and install lots of surface area, but if it is covered with detritus or calcium, the heat transfer will dramatically drop off and the condensing pressure will rise, negating any benefit of the additional CAPEX spent.

As the condenser is the last heat exchanger to see the effects of any inefficiencies in the refrigeration system or process, from an energy efficiency perspective, I have always considered it as being the most important piece of equipment in a refrigeration system when considering both design as well as maintenance.

As with evaporators, condensers are heat exchangers and as such should be kept clean.

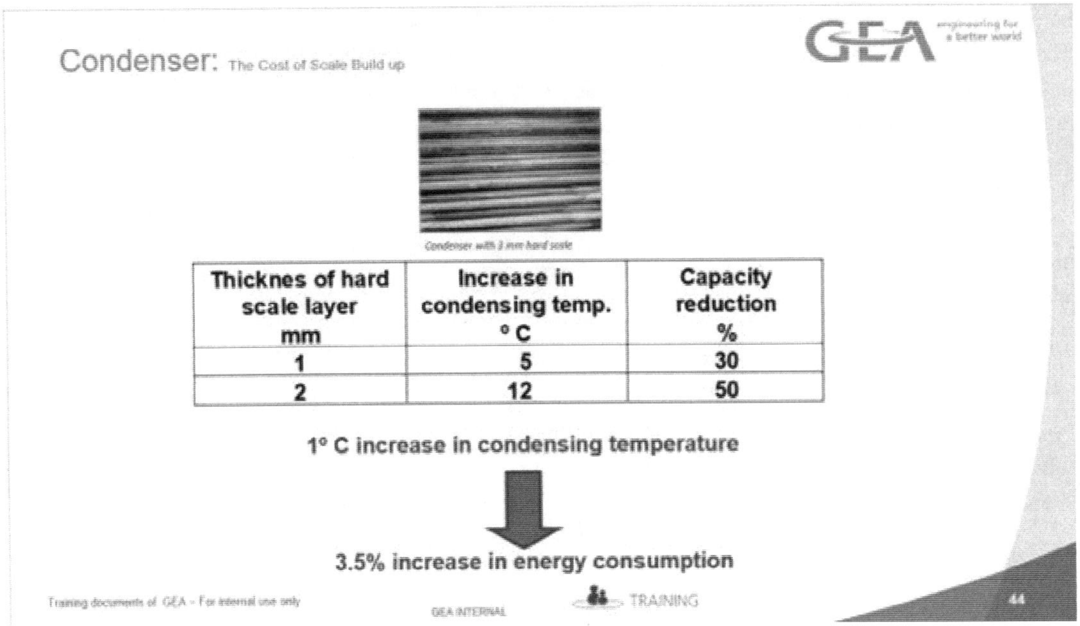


| Thicknes of hard scale layer mm | Increase in condensing temp. °C | Capacity reduction % |
|---|---|---|
| 1 | 5 | 30 |
| 2 | 12 | 50 |

**Figure 4: The effects of scale build up on the coil of an evaporative condenser**

If the coil on the evaporative condenser is covered with calcium deposits for example, or the water distribution system is not correct, the effect will be this heat will be rejected to the ambient air with a higher temperature difference than design, and for every 1k, the condensing saturated temperature raises because of this "insulation", the compressor draws approximately 3.5% more power for the given cooling duty. Again, it's no good installing expensive heat exchangers and not maintaining them appropriately.

It is common practice these days to control condensers using "floating condensing pressure control", effectively controlling the optimum temperature difference between the ambient temperature and the saturated condensing temperature, rather than at a fixed point. This requires additional control and sensors but can give good returns on investment if designed properly. Consideration must also be given to items such as expansion valves, discharge and hot gas supply systems and oil separators.

It's a common misbelief that lowering the condensing pressure gives less heat for defrosting for example, but this is simply inaccurate. The heat given up per kg (latent heat of condensation) condensed is greater at a low pressure than a higher one. The issue is, as the pressure decreases in the discharge system, the specific volume of vapour increases dramatically. For example, the difference in a specific volume of the vapour at +35°C and +15°C on ammonia is appx. 45% greater at the lower pressure.

Therefore, if you try and lower the condensing pressure to these levels and the system hasn't been designed for it, the velocities in all the pipes and oil separators will increase dramatically, so a point where defrosts take longer and oil is blown out through inadequately sized oil separators and coalescent filters.

An expansion valve orifice can merely be taken up one size to cope with the lower condensing pressure.

Remember, 50% of the year is night, and the other 50% is divided between seasons. For example, the effect on a compressor running at -10°C/+32°C during summer and -10°C/+15°C during winter is a COP of 4.13 and 7.1 on the same compressor, so a 42% efficiency increase. A little more CAPEX spent on these items can achieve a much better performance, and obviously, the condenser must also be kept clean throughout the year.

## 4. Conclusion

Refrigeration is a huge user of energy for the end user, and if you are a cold store owner it is not doubt by far your largest consumer.

There are many items of design (some covered above) which can raise the performance of these systems, and although they may involve increased CAPEX expenditure, they have quite good returns on investment. Although I was asked to argue the point for spending funds on efficient design, and I hope I have given some examples of this, these examples can be quickly negated by poor maintenance, so I fear the answer to the question, "Where should I spend my cash", it's always going to be - both.

I would however like to point out, you hopefully only spend CAPEX once, and if well-designed and maintained, our systems will outlast us. When spending a little extra for those few hours for your refrigeration service provider to check the correct operation of the system, it's arguably money well spent and is relatively inexpensive.

I always wonder why people spend so much per hour when they deliver their car into their annual service when the engine only ran typically for at best 1 hour per day doing the school run, where the "Engineer" only plugs it into a computer to tell you to top up with oil and perhaps change the filter, where the compressor which runs 24/7 and a system that makes you money gets quibbled over because the hourly rate for a real Refrigeration Engineer is half as much. Maybe that's a good point for discussion...

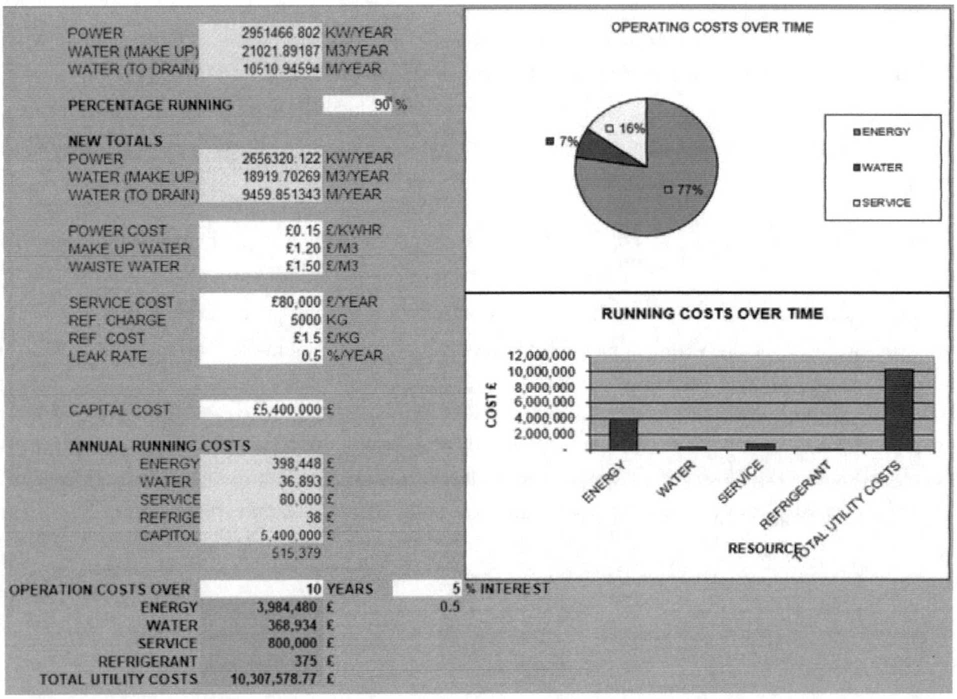


**Figure 5: Life cycle cost calculation over 10 years based on 1MW -30ºC, two stage system with 20 Evaporators and evaporative condensers, with a "Guesstimated" CAPEX.**

Figure 5 above details the results of a life cycle cost calculation with some estimations related to various costs, based on a 1,000kW design refrigeration load on a cold store with 20 evaporators operating with a suction pressure of -30°C, summer condensing condition of +32°C, with a two-stage, pump overfed system.

When one looks at the overall expenditure, excluding the CAPEX, the running costs are mostly made up of energy, and only 16% is accounted for with maintenance costs. If we said that half of the maintenance is components and the rest is labour, you could nearly double the amount of labour at the same cost of a 10% efficiency increase. With a correctly trained/experienced refrigeration engineer, you would only need to spend 10-20% more in hours if properly utilised to easily recover these additional costs many times over.

Proper maintenance should enhance the energy efficiency of your refrigeration system and is an inexpensive way of going about it, and now I am not sure if I am on the right side of the table....

## References

Felix Webber. Thermofin GmbH;

Ken Picking. BAC

## DISCUSSION REPORT

**Dermot Cotter**, chairing, thanked the authors for a very interesting paper.
I'll start the questions with asking - I'm a client and I have one person to install my system and someone else to maintain it, why should I bother spending money on maintaining it? It's working, it does the job. Why should I spend more money Lisa?

**Lisa Pogson** – Let's compare it to a vehicle that customers have some knowledge of as people don't have a lot of knowledge of air conditioning or refrigeration. If you have the best car in the world but never get it serviced i.e. water, oil and filter checks then it will run but at some point, it will start stuttering and sometimes you notice that too late. The example I give to customers is if you maintain your equipment and do the basic checks then it will probably run a lot longer than you would expect it to. There's typically a 10-15 year life cycle on air conditioning systems in offices but we see some great systems that have been properly maintained lasting for 25 years, sometimes more. When you are spending money on a system and having it designed well then you also want it to last because design costs have a 10-year payback and when you can 15 or 20 years it is even better and that is important to me as a consumer myself.

**Dermot Cotter** – So to keep something operating with maintenance and to make it very efficient with maintenance how much money would you need to spend?
**Lisa Pogson** – There's often a discussion between doing basic checks and 'good' maintenance. Basic checks are checking things are switching on and off properly, keeping things clean and tidy. The basic work is what we recommend and anything else is totally down to what the client is prepared to pay for. If they are willing to spend more or there are people on site who can monitor the system then that's great but they must do basic checks and work up from there.
**Robert Unsworth** – I'll counter that. Obviously, I work for a compressor manufacturer and from a compressor manufacturer's point of view, most refrigeration systems have a spare compressor sitting there anyway. A screw compressor should run for 50,000 hours before you need to touch it. As long as it's a well-designed system you should have oil return that is clean and filtered. If you can see particles in oil then you're already too late - so what's the point in maintaining it? Design it properly so you don't have these issues to begin with. The greatest cost to the client is on energy running costs so if you have limited budget put money into efficiency. If you have bigger surfaces in your condenser and your air coolers then they won't self-clean but dirt on the cooler will not have as great effect. To go back to the car, I don't need an engineer to clean it and I've had cars where the engine stops working and I just replace the vehicle every 10 years or so anyway.

**Lisa Pogson** – I'll counter that as well. From a sustainability point of view, we don't just want to keep throwing systems out and putting new ones in. I do agree that it's not just all about cleaning. In the systems that we work on, a lot of comfort cooling, which customers use as heating and cooling these days, if something is broken and not checked people just leave it running. We find a lot of the time that they are not setting it up properly and it's not running efficiently. I know of a building that had a very energy-efficient system that was a very new water-cooled system and they really wanted it to work properly but what they had on site were people that were leaving it on continuously, they didn't check when it was running, compressors were going down and we found over time that a system that should have run for a long time and should have been really energy efficient and should have saved them significant amounts of money on energy was doing the exact opposite and they were getting charged lots of money because they didn't have anyone on site to check it. They had someone in the plant room adjusting it who didn't understand the system. So it's having things checked and monitored over time. What gets measured gets managed and the monitoring of it from a maintenance point of view is very important.

**Robert Unsworth** – At the end of the day a lot of the things we are talking about could be covered by controls. Automated systems and monitoring of refrigeration systems are becoming more user friendly and the controls to keep the system running optimally are built into the control system, as well as the monitoring, so you don't need an engineer to check things because it is covered by the control philosophy and the monitoring systems.

**Dermot Cotter** – I've looked at many systems and I find things like EC fans that are run as fixed speed fans in evaporators, no oil return systems - you're draining out oil and putting oil in. What is the cost difference of putting in a really efficient system in comparison with a basic refrigeration system? In other words equivalent to a D class system as opposed to an A+ system.

**Robert Unsworth** – Maybe 10% and that's nothing relative to the lifetime of a refrigeration system. You only spend that money once generally and rather than spending £5 million upfront you spend £5.5 million on pieces of equipment that will make the system run more efficiently and make them run efficiently for longer. So you might be buying a bigger reciprocating compressor that runs slower, which not only saves you energy, so COP goes up, but it also extends the life of the compressor. It's like car engines where you don't just fix them at fixed hours, you fix them when it tells you to and that's a function of the speed of the compressor and the temperature difference across it so if you use things like floating head pressure control then you extend service intervals and your cost falls. To do those things, to buy a screw compressor with an economiser with variable speed and all the auto VI function on the screw compressor proportionally against your capital expenditure it's relatively nothing cost-wise. You spend it once and these items are the things that save you the money in your operating costs. If you know what you're doing and look at an industrial refrigeration system then you can tweak things manually and make such huge benefits - but that is because the client hasn't bought the items that would have done those tweaks automatically in the first place. If they had spent the money on those things when they installed the system then they might have spent 10-15% more on the project but the return on the investment would be 18 months. Unfortunately, that's not what most clients look at. How much will I have to spend now and when will I get my money back? Although these days with the increased cost of electricity it has meant that these items that make refrigeration systems run more efficiently just have a shorter return on investment so doing the right thing costs less at the moment.

**Dermot Cotter** – Do you have any data about replacing older refrigeration systems and putting in efficient systems, what effect does it have on energy savings and how can it reduce your maintenance cost?

**Robert Unsworth** – No. The compressors that I tend to put in have a typical life of 50 years so they don't tend to get replaced. If you have a chill system with a screw compressor then it may get swapped out for a reciprocating compressor when the compressor is up for renewal because thermodynamically it makes sense and they'll get an efficiency gain out of that. It's always based on return on investment. If you are going to put an energy-efficient system in then you should do it from day one otherwise you end up spending more money to do it afterwards. You can always do it but is it capitally viable to do it at that point?

**Dermot Cotter** – Lisa, when looking at air conditioning systems do you find energy meters in place so customers know what power they are using? Does anyone on site know what is happening with them?

**Lisa Pogson** – We have found it a lot more in the last year, as the energy costs go up. To be honest, it is something that we have talked about for years, and that is something people should be monitoring because it goes into those return on investment figures. Customers put systems in for 10 years in offices or laboratories or education settings and they are usually looking at value engineering things down. They aren't looking to spend another 10%. Although in this last year they are starting to put a lot more energy monitoring in place and looking at setting things up properly.

**Robert Unsworth** – I'll just counter that because just measuring power is not an indication of what you are doing. If I have a 250 kilowatt motor it doesn't mean that my compressor is doing X kilowatts worth of cooling. You could have an inefficient compressor doing a little bit of cooling or a very efficient compressor doing a lot of cooling. You have to able to measure the cooling side as well and that is actually quite complicated.

**Dermot Cotter** – A question for you Robert. I hear about commissioning a plant properly but is the maintenance not a continuation of the commissioning?
**Robert Unsworth** – Yes but, again, depending on how much money you want to spend when you install the system. You should commission it properly and design it properly. Quite often I look at refrigeration systems in other countries and find that clients have their own refrigeration engineers (most clients in the UK have outsourced that) and those plant rooms tend to be clean and well-maintained. However, a lot of them aren't designed well or controlled well because whoever installed them didn't have the skills at that time but they are all very clean but not running very efficiently.

**Dermot Cotter** – So, if I'm an end user saying please put in an energy-efficient refrigeration system for me, how do I know that you will actually do it? From my own background, I see lots of (even brand-new systems) that are not actually efficient. They're 15-25% energy inefficient compared with where they should be. How do you tell, how do you get a customer to spend their money if everybody else is just selling cheap inefficient systems?
**Robert Unsworth** – During their specification, they should be specifying how efficient it is going to be and then it should be monitored afterwards. If you are putting in a bigger system with a monitoring system then it should be measured and the client sets KPIs and we should be measured against them so you have to measure the performance at full load and at part load and over periods of time. They then become your KPIs and your refrigeration system should then operate within the window of that KPI.

**Dermot Cotter** – Lisa do your clients use KPI for system performance in air conditioning?
**Lisa Pogson** – No, and what is important for me is that the client stays a client through to the maintenance, because quite often we are installing systems for landlords who pass it on to an occupier and so they are not bothered about running costs or operation. I'm sure the system is set to operate properly but they are not monitoring it further on. What I find really interesting is when a client is having the system specified and designed and then they are going to carry on using it themselves and carry on paying for the maintenance because that is where you usually find that the clients will be quite good about getting the maintenance done. One of the things in the paper that Rob wrote, is that the capital expenditure only happens once and the maintenance cost is ongoing. But I think it's really important that you have to keep going back and checking on operation. What we are finding now is that you have got to really engage with the customers. Customers are coming from all those angles. They're coming from cost, they're coming from compliance, they're coming from sustainability but where is their main focus? Over time it might be that they have to meet carbon targets and even in London now you have to do certain things to get your rates reduced. That's having an effect as well as the higher energy costs so there's certain factors that you've got to be thinking about over time.

**Dermot Cotter** –What kind of policy should the government be introducing to encourage better maintenance and also what policy should they be introducing to ensure better, energy efficient

equipment is installed rather than what I would consider 'standard' stuff that is put in at the moment?
**Lisa Pogson** – Encouraging people via legislation to get systems checked. There is no policing of current requirements. If there is no checking of anything and you don't monitor it then people don't manage it. So having somebody that is going to police that. The other part of it is the sustainability, the long term implications of the equipment that you are putting in and encouraging people to think twice and not just throw things out and replace with a new system.

**Dermot Cotter** – What legislation should the government bring in so that people are putting in efficient refrigeration systems and not just your standard?
**Robert Unsworth** – Actually I thought there was legislation on the way. At the end of the day this all goes back to carbon and decarbonisation. Refrigeration is a huge energy user globally and for the small amount of capital expenditure increase to put an efficient refrigeration system in, whether that is done through legislation or there are grants and incentives for certain pieces of equipment. You can put an efficient piece of equipment in but if it's not controlled correctly then it won't be very efficient. KPIs should be put in place with regards to refrigeration systems and that would cover an official COP for a certain refrigeration system and then there should be KPIs for its operation throughout the year. That could be set in place and you would have to fall within that. But the argument is that if you fall outside of that KPI band then who gets penalised? Is it the contractor or the maintenance company or is it the client themselves? I suppose it is the end users responsibility and if the system operated well at the point at which it was commissioned and then didn't 3 years later arguably that is because of poor maintenance or wear or whatever.

**Dermot Cotter –** What I am seeing from the comments coming in is that there are gaps between maintenance and design and what is highlighted here is that clients just don't understand where they should spend the money and where they should not and there's no guidance out there.

A poll was then conducted with the audience on whether Maintenance or Efficient Design is more important. The results were: Maintenance – 25% Energy efficient design – 75%

**Robert Unsworth** – Thank you very much but all of you that voted for me arguing for Design, I'm now going to tell you that I believe both are equally important.

**Dermot Cotter** – It is a very awkward one for anyone arguing it because everybody needs both. Lisa from your point of view what do you actually think?

**Lisa Pogson** – To be honest, before I came in today I was really forcing the point of maintenance but I did have a chat with our head of design and he put me right that if you don't spend the money up front then it's never going to run right and it's never going to be working the way it should. But I do think that commissioning it properly and running it well afterwards, making sure the controls are in place is really important so I accept 50/50.

**Dermot Cotter** – The one thing I hear from the two of you is that you both think from a maintenance point of view we need legislation because if we carry on as things are, then people are not going to do it. Is that the picture I am getting?

**Robert Unsworth** –If it was your car and you were doing 600,000 miles a year in it, then you would look after it for your wellbeing and the wellbeing of your children relied on it. It doesn't matter if a system has lots of surface area and you are controlling everything properly if it is full of oil and covered or blocked. It has a direct influence on the energy consumption of the plant which is by far the largest chunk of expenditure. So you must spend your money from day one in the capital expenditure and your correctly designed refrigeration system and that should have KPIs put in place to make sure it operates well and as it was designed and it should be looked after. These are expensive pieces of equipment and they're going to be there longer than us so they need to be maintained properly.

**Dermot Cotter** – Thank you both for delivering a good debate and putting a different slant on these important issues.